Principles of Brewing Science

A Study of Serious Brewing Issues
SECOND EDITION

GEORGE FIX, Ph. D.

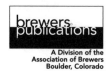

brewers
publications

A Division of the
Association of Brewers
Boulder, Colorado

Brewers Publications
PO Box 1679, Boulder CO 80306-1679
Tel: 303-447-0816   Fax: 303-447-2825
www.beertown.org

ISBN 0-937381-74-8

Library of Congress Cataloging-in-Publication Data

Fix, George J., 1939—
    Principles of brewing science : a study of serious brewing issues
    George Fix. — 2nd ed.
        p.   cm.
    Includes bibliographical references and index.
    ISBN 0-937381-74-8 (pbk.)
    1. Brewing.   I. Title
TP570.F58 1999
663'.3—dc21   99-36534
                                                  CIP

Printed in the United States of America

10   9 8

Contents

Figures

Tables

ACKNOWLEDGMENTS

Many people have contributed to the production of this book. Foremost is my wife, Laurie Ann Fix. She is my companion, confidant, best friend, and love of my life. Without her, nothing would have been possible.

I am also grateful for the significant input of Scott Bickman, a technical reviewer of this book. Dr. Bickman is a scienist of accomplishment and depth. He also has a striking ability to distinguish between issues that are crucial in brewing from those which are merely theoretically possible. I would also like to acknowledge the many comments, corrections, and abservations made by Dr. Rick Wood. In fact, the contributions of Drs. Bickman and Wood were integral to the completion of this book. Nevertheless, I take full responsibility for any problems associated with the material in this book.

The late Gilbert Straub was my brewing mentor, and his "Weltweisheit" of brewing guides me to this very day. Under his leadership, the tiny Straub Company sailed through the post- World War II era at full capacity, while other small operations were closing. Even in the early 1980s when his was only one of three commercial breweries with a capacity under 25,000 bbls., it was clear that the Straub Brewing Company was going to be around for the long haul. There is a lesson for small commercial breweries in this history.

I am also grateful to Brewers Publications and my publisher Toni Knapp, for guidance in the preparation of this book. The technical editor, Mary Eberle, made significant and important contributions to the accuracy of the material. This book could never have been published without their editorial management.

Finally, I would like to cite all the brewers, who are too numerous to mention, including those who have participated on the Professional Tasting Panel of the Great American Beer Festival, members of the MBAA and ASBC, and the remarkable and stimulating discussions that take place at the meetings and conferences. Also to be acknowledged is an even longer list of homebrewers. Over the years homebrewing conferences, competitions, and electronic forums like Homebrewers Digest have added greatly to my understanding of brewing.

Introduction

The great Viennese composer, Anton Webern (1883-1945), once remarked, "All art, all music is based on laws" (Bailey, 1998). Webern was not denying creativity and originality. Indeed, he was one of the twentieth century's most innovative musical figures. Rather, he saw fundamental principles as a guide to lead him down the path to the new music he wanted to write, and to understanding the past masters he so revered.

Brewers are in a remarkably similar situation. Creativity and originality are shared values for all of us. What separates this highly heterogeneous group are our objectives. This, too, has an analogy in music. For example, the only thing Webern and Johann Strauss appear to have had in common was that their major works were composed in or near Vienna. Strauss sought popular success, and Webern artistic success. Yet both used the same principles of harmony and counterpoint, albeit in radically different ways! Put differently, things happen for a reason, and fundamental laws are our best guide to understanding the consequences of our actions.

The intriguing situation about brewing, on the other hand, is that many mechanisms are theoretically possible, and the real key to success is the ability to identify those that are genuinely relevant in any particular situation.

I would like to identify what I feel is a rank ordering of material in terms of relevance. Relevance in brewing exists on three levels—primary, secondary, and tertiary. Primary effects are crucial, and every serious brewer needs to completely master these. Dealing with secondary and tertiary effects amounts to varying degrees of fine tuning.

Fermentation. As long as beer flavor is the final arbiter of what is desirable, the fermentation of beer will be the area in brewing of greatest practical importance. Moreover, this is as true for conventional beers brewed with conventional techniques as it is for specialty beers.

A mastery of the various metabolic pathways is highly recommended for even beginning brewers. The major pathway, often called the Embden-Meyerhof-Parras pathway, consists of Glycolysis, where each molecule of glucose is transformed to two molecules of ethanol and two molecules of carbon dioxide. A strong case can be made that the first part of this pathway is the most crucial. It is certainly the part of the fermentation where we as brewers can exert the greatest control. In this regard, consider the following four points.

First, the yeast used must be in excellent physiological condition. The yeast membrane controls what enters and leaves the yeast cell, and its viability is a necessary condition for a satisfactory start of the fermentation. Also important is for the yeast to have an adequate food supply within their cells.

The second and closely related issue is to add an adequate number of yeast cells. At the start of the fermentation more yeast is definitely better than less, however, too much is not good either. Indeed, one of the major themes that permeate this book is that successful brewing is one where the correct balance is struck in all issues. Fortunately, there are elementary procedures for measuring viability food reserves and cell counts.

The third point associated with the initial period is the indispensable role that oxygenation of chilled wort plays as a yeast nutrient. The most important aspect of exogenous oxygen is associated with yeast synthesis of sterols and unsaturated fatty acids, a crucial step in the maintenance of yeast viability. Yet there is more, and the inevitable conclusion one reaches is that wort oxygenation can never be omitted by quality conscious brewers.

The fourth and final point about the start of the fermentation relates to wort composition, and the vital role played by key wort constituents such as nitrogen from amino acids.

Other parts of beer fermentation that fall into the primary relevance category are the various "minor" metabolic pathways. These produce such flavor active products as esters, fusel alcohols, diacetyl, and phenols, to cite a few examples. The term *minor* is used for these pathways because only a miniscule amount of products are produced in them, typically less than 1 mg per liter. This is in striking contrast to the main pathway where over 40,000 mg per liter of ethanol is produced in beers of normal strength. Yet, what I find most fascinating about brewing is that minor products can influence beers flavors in ways that far exceed their concentrations. Put differently, the minor pathways can become major once the quality of beer flavor is put on the table.

Three points about the minor pathways include, first, the role played by individual yeast strains. Given the various metabolic options, it is not surprising that different strains display different propensities for divertissements from the major pathway. This defines the strain's basic flavor signature, and is the brewer's most reliable criterion for strain selection.

Second, and of equal importance, is, again, the need for the yeast cells used to be in excellent physiological condition. A version of Murphy's law states that yeasts which have not been afforded the best of care will start to do things they should not do, and stop doing those things we want them to do. Unfortunately, there are plenty of opportunities for yeast to misbehave.

Last, is the issue associated with noncultured yeast and/or beer-spoiling bacteria. There is an old brewer's saying that large infections are not a problem in brewing, but the small ones are killers. Indeed, the former are easy to detect, and in most circumstances, easy to reverse engines to identify the culprits. Dealing with minor infections is considerably more subtle. For example, my experience has been that most complaints about a beer's lack of maltiness is not due to the grains used or the way they were processed, but rather to minor infections.

There are many other examples along these lines that display analogous effects. No brewery, amateur or commercial, is sterile in the sense of being free of all hostile microbes. Moreover, attempts to render them sterile would lead to a situation where the cure is worse than the disease. Yet, it is important to know whether we fall into the low and trivial category or something less desirable. Fortunately, there are elementary procedures for determining this condition. On my wish list is that, you will use one of the many options for determining undesirable microbial levels.

Oxidation. The deleterious effects of oxygen uptake at any point in the brewing cycle is well documented. The only exception to this is the oxygen introduced at the start of the fermentation. It is true that there is a considerable variation among beer drinkers both with respect to their ability to detect oxidized flavors, and with respect to their acceptance of these notes. Yet the track record is clear in both amateur and commercial brewing: Consistently successful brewers are invariably the ones who operate low oxygen systems.

For most of the twentieth century, attention was focused on oxidation occurring after the end of the fermentation—so-called cold side aeration (CSA). Concerns about CSA are well founded since there are a number of relevant mechanisms, all of which are destructive to beer flavor. However, it has been clear for a long time that there is a lot more to oxidation than CSA. Indeed, this became particularly apparent in the 1980s when modern double pre-evacuation bottle fillers were put in place. This reduced package air content to miniscule levels, yet problems with oxidation continued, particularly with long chained unsaturated aldehydes like trans-2-nonenal (T-2-N). These compounds have a very low flavor threshold, and contribute the classic stale, papery, cardboard tones.

Additional research showed clearly that these staling aldehydes cannot arise from CSA, and must come from other causes. Attention quickly focused to wort production, and what is usually called hot side aeration (HSA). Two new mechanisms were discovered.

The first is the "herbstoffe effect." *Herbstoffe* roughly means grain astringent in German. In this mechanism, malt constituents, most notably phenols, are oxidized during wort production. They then spill over into the finished beer creating the herbstoffe notes. It is not uncommon for metallic notes to accompany these flavors. What is particularly pernicious about this effect is that it is often misdiagnosed, and not associated with oxidation. This of course makes the resolution of the problem more difficult.

The second mechanism is usually called "oxidation without molecular oxygen." Here oxidized grain constituents, most notably melanoidins, spill over into the finished beer and in turn oxidize alcohols. The products from this mechanism are similar to those associated with CSA.

These findings led to a revolution in brewhouse technology, and by the 1990s the low oxygen brewhouse became common. This technology, along with low oxygen bottle fillers, did improve beer stability with respect to oxidation. However, and alas, neither totally solved the T-2-N problem.

New research focused on unsaturated fatty acids (UFA) in wort as T-2-N precursors. Part of this was already well established. For example, the nonenzymatic oxidation of UFAs during wort production was a well-documented part of HSA. However, new mechanisms have been discovered as well. What is particularly fascinating about them is that they are induced by malt enzymes, and can take place even if HSA is absent. These results confirmed the long held suspicion that malt and malting were a factor in T-2-N formation along with HSA.

SECONDARY RELEVANCE

The term "secondary effect" may strike some as being the same as unimportant. In brewing this is anything but the case. For example, in amateur brewing a mastery of the secondary effects is usually worth at least five points in competition. Given the high quality of home-brew being produced, this could very well be the difference between first and last place. In commercial brewing the situation is even more serious. Indeed, there is considerable evidence which indicates that brand loyalty among beer drinkers is by and large determined by secondary effects.

Malt. Some may take issue with this classification, for malt is certainly the soul of beer. Yet, given the very high quality of malt available to brewers, it is not the flashpoint that it was in the 1980s and earlier. Sometimes I have the feeling that it is much harder to make a bad choice of grains than a good one! What this means is that as far as grains are concerned the choices range from good to better to best. Nevertheless, the quality conscious brewer will always want the best.

Maillard products. The presence and desirability of so-called Maillard products is an important issue associated with malt. These are formed during the hot part of malt processing, i.e., during grain drying (kilning) in malting, during mashing, and finally during wort boiling. The Maillard products exist in beer in three basic forms. The first are the elementary melanoids, which are universally prized for their "fine malt" flavor characteristics. The intermediate and complex Maillard products bring special flavors to beer that may or may not be desirable.

The major conclusion to be gleaned from the general Maillard theory is that it is advantageous for brewers to give serious attention to the overall heat treatment that has taken place during malting, mashing, and wort boiling. Different brewers will undoubtedly come to different conclusions about the acceptability of the Maillard products, even in the context of a particular beer style. Nevertheless, what is discernibly true for all is that the effects due to Maillard reactions legitimately belong to the secondary relevance category, and therefore deserves serious attention.

Hops and hop chemistry. There are two practical conclusions that I hope you will get from this material. The first is the significant differences in the chemical composition of hop varieties, and how this fundamentally effects their flavoring potential. The second point concerns the stability of hop flavor and aroma. There is nothing more frustrating than to have a beautiful hop taste and smell in fresh beer, only for both to disappear as the beer ages. Moreover, the speed at which this process can occur shows that it cannot be blamed entirely on oxidation.

TERTIARY RELEVANCE

Beer clarification. It has been my experience that with the high quality of malt and yeast available, beers should naturally clarify without assistance. A point that I stress is that if this is not the case, then there are invariably problems elsewhere which are the culprit. The best course of action is not to use clarification techniques, but rather to address the fundamental problems that created the haze in the first place. Nevertheless, there are circumstances where it is desirable to polish up a beer, and render it stable with respect to temperature changes and mechanical abuse. One approach uses additives, and the other uses non-additive approaches such as ice stabilization.

Water chemistry and water treatment. The central thesis put forward is that there are two crucial aspects associated with brewing water. First, it should be free of impurities, and have a pleasant taste. The other consideration is that the water be of such a composition that the mash and wort pH fall into a desirable range.

Minerals are sometimes added to achieve various stylistic effects. A fundamental question is whether extensive use of mineral salts is the best way to influence beer flavors. The answer to this will by and large depend on the brewer's motivations, but I for one would rather depend on the basic materials (i. e. , malt, hops, and yeast). I cite

---

examples illustrating this along with supporting data from highly regarded commercial beers.

Finally, having discussed at great length the importance of relevance, I would like to point out that this book concludes with a discussion of the dispensing of beer, an appealing topic not normally found in books on brewing science. One reason for its inclusion has been my observation that many homebrewers have trouble getting CO_2 levels right when they first start working with home draft units. The same also applies to inexperienced operators of commercial draft systems. There is really no reason for this since there are some easy to understand gas laws that can guide one down the proper path.

As I quoted Anton Webern in the beginning, "All art, all music, is based on laws." And so it is in brewing.

Malting and Mashing 1

Malting and mashing should be considered as closely related aspects of the same process. In effect, both involve the breakdown of key grain constituents—most notably starch and proteins—into simpler units. Brewing yeasts are an elementary type of microbe in the sense that they can only metabolize simple sugars and proteins. Hence, degradation that occurs in malting and mashing is crucial for beer. Water plays a role in both malting and mashing. The former starts with steeping, a process in which grains are immersed in water. The resulting moisture uptake into the barley kernels marks the beginning of the modification process. As a consequence, chapter 1 starts with a discussion of brewing water, which is usually called brewing liquor. Waters of highly diverse composition have been successfully used in brewing, and even with a particular beer style, there are a number of reasonable options to adjust the water composition. Therefore, the fundamental issues are discussed to provide brewers a base from which they can decide among the diverse water-treatment options.

Logically the next subject is a review of the key grain constituents, namely, carbohydrates and proteins. Also included are other grain constituents, most notably phenolic and sulfur compounds, which play a major role in finished beer flavor. Acquiring a good grasp of the structural properties of these compounds can be of great value to brewers when deciding on the type of brewing program that will be used for a given beer style.

In the rest of this chapter, the point of view taken is that both malting and mashing are in effect major enzyme-induced transformations in the presence of water. This theme will reemerge throughout this book, and it is not farfetched to see most of brewing as a series of enzyme-induced transformations. Water-borne minerals affect enzymatic activity in the mash. Also important to malting and, to a lesser extent, to mashing, are the heat-induced Maillard (i.e., the brown-pigment-producing) reactions. As these are topics of great practical importance, they are taken up again in chapter 2.

BREWING LIQUOR

Most beers are between 90% and 94% water; hence, it is by a wide margin the most abundant beer constituent. Certain properties of water are especially relevant to beer. The primary application of a discussion of these properties is to help the brewer decide how a given source of raw water can be transformed into "brewing liquor," i.e., water that has been suitably treated so that it can be advantageously used in the brewing process for a particular type of beer.

Water is an excellent solvent, a property that is crucial to brewing. Otherwise, the brewer's efforts to dissolve grain constituents in the mash and hop constituents during wort boiling would be for naught. The key to the solvency of water lies in its *polarity*. In particular, a water molecule (H_2O) can be represented as follows:

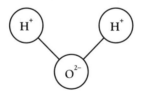

Water molecule

This molecule is electrically neutral because the oxygen ion O^{2-} has two negative charges, whereas each of the two hydrogen ions H^+ has one positive charge. Nevertheless, near the oxygen vertex of the H_2O molecule, the overall-neutral molecule has a partial negative charge. The hydrogen side of the H_2O molecule, on the other hand, is positively charged. The existence of a negative region and a positive region on an electrically neutral molecule is called *polarity*.

To see how this works, consider calcium chloride ($CaCl_2$), a salt that is widely used in brewing to treat water. Once added, this salt tends to (partially) dissociate into three ions: one Ca^{2+} ion and two Cl^- ions. These ions are held in solution by the polar charges of the H_2O molecule as follows:

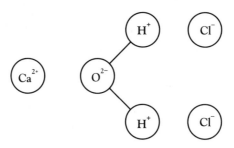

Calcium and chloride ions in solution

In this book, dissociation is represented by using the following notation, for example:

$$CaCl_2 \leftrightarrow Ca^{2+} + 2\,Cl^- . \tag{1.1}$$

The double arrows indicate that the reaction can go both ways; i.e., added calcium chloride will dissociate into ions; however, there could also be undissociated calcium chloride molecules. Calcium sulfate (sometimes called gypsum; $CaSO_4$) is another salt used to treat brewing water. The dissociation reaction for it is

$$CaSO_4 \leftrightarrow Ca^{2+} + SO_4^{2-} , \tag{1.2}$$

where SO_4^{2-} is the sulfate ion. The primary source of magnesium ion, Mg^{2+}, in brewing water typically comes from magnesium sulfate (sometimes called Epsom salt; $MgSO_4$) via the dissociation reaction

$$MgSO_4 \leftrightarrow Mg^{2+} + SO_4^{2-} . \tag{1.3}$$

Calcium carbonate ($CaCO_3$) is often used to treat water for dark beers. It partially dissociates into calcium and carbonate ions:

$$CaCO_3 \leftrightarrow Ca^{2+} + CO_3^{2-}. \tag{1.4}$$

A different situation is presented by common table salt (NaCl), which is also used in brewing. It tends to completely dissociate, and hence the dissociation reaction for it is written with only a single arrow that points toward the ionic side of the reaction:

$$NaCl \rightarrow Na^+ + Cl^-. \tag{1.5}$$

In fact, most water-treatment procedures tend to promote full dissociation (by stirring and other means) of any salt that is added to the water. Thus, it can be assumed, unless otherwise noted, that the right side of reactions 1.1 to 1.4 is favored. Hence these dissociation reactions can be represented as follows:

$$CaCl_2 \rightarrow Ca^{2+} + 2\,Cl^- \tag{1.6}$$
$$CaSO_4 \rightarrow Ca^{2+} + SO_4^{2-}$$
$$MgSO_4 \rightarrow Mg^{2+} + SO_4^{2-}$$
$$CaCO_3 \rightarrow Ca^{2+} + CO_3^{2-}$$

Water-borne ions play an important role in brewing, and calcium ions are used as an example. These ions react with malt phosphates (primarily K_2HPO_4) and hence decrease pH (i.e., make the water more acidic; pH is a measure of the acidity—lower values indicate a more acidic solution compared to higher values) via

$$3Ca^{2+} + 2HPO_4^{2-} \leftrightarrow 2H^+ + Ca_3(PO_4)_2. \tag{1.7}$$

The second term on the right is a calcium phosphate and is essentially insoluble in wort. Kolbach's (1960) studies clearly show that this reaction does not go to completion, and his data indicate that 5 calcium equivalents are needed to produce 2 equivalents of hydrogen. Moll (1994), using improved instrumentation, found that approximately 7

---

calcium equivalents are needed. This 7:2 ratio is much more than the 3:2 ratio expected from a 100% utilization in reaction 1.7. The implications of this finding for water treatment are discussed later in this chapter.

Calcium ions also tend to afford thermal protection for mash enzymes (Comrie, 1967). In addition, they continue to interact with malt phosphate during wort boiling, and the ongoing reaction between calcium and phosphate is the primary reason that the pH decreases in the kettle boil. Calcium ions also tend to inhibit color formation during the boil and facilitate protein coagulation. Finally, calcium ions also influence beer fermentation. For example, they favorably affect yeast flocculation and beer clarification during maturation (Harrison et al., 1963; Saltukoglu and Slaughter, 1983; Taylor, 1990).

Magnesium also plays a role in mash acidification by a mechanism similar to that described in reaction 1.7. However, these effects are small and are usually ignored. The major role magnesium plays is in the fermentation where magnesium is an important cofactor in various metabolic activities. Actually, malt supplies sufficient magnesium for this purpose, and in addition, at too high of a concentration, magnesium ions will contribute a harsh bitterness (Krauss et al., 1983). As a consequence, magnesium salts are rarely used for treating water in modern brewing.

Carbonate (CO_3^{2-}) and bicarbonate (HCO_3^-) ions are very important to wort production. They work in the manner opposite to Ca^{2+} ions in the sense that they tend to increase pH by binding H^+ ions to other compounds:

$$CO_3^{2-} + H^+ \leftrightarrow HCO_3^- \tag{1.8}$$
$$HCO_3^- + H^+ \leftrightarrow H_2CO_3 .$$

In water, these reactions are equivalent to the liberation of hydroxyl ions (OH^-):

$$CO_3^{2-} + H_2O \leftrightarrow OH^- + HCO_3^- \tag{1.9}$$
$$HCO_3^- + H_2O \leftrightarrow OH^- + H_2CO_3 .$$

The last compound, carbonic acid (H_2CO_3), is neutral and thus does not change the pH of wort. In fact, it decomposes to carbon dioxide (CO_2) via

$$H_2CO_3 \leftrightarrow H_2O + CO_2 ,$$

(1.10)

and the CO_2 for the most part evolves as a gas under normal mashing conditions. The proportions of reactants and products that reactions 1.8 to 1.10 yield depend on the pH of the media, as can be seen in Table 1.1.

A widely accepted rule in brewing is to have calcium concentrations of at least 50 mg/L, and values in the range of 100–150 mg/L are very common. Such a range of concentration is denoted

$$<Ca^{2+}> = 50-150 \text{ mg/L} ,$$

(1.11)

where the angle brackets < > indicate ionic concentration in mg/L. In most practical brewing situations, the available water is calcium deficient. To achieve the criterion specified in equation 1.11, calcium-based salts are typically added. Calcium chloride ($CaCl_2$) is widely used for this purpose. It usually contains water of hydration, and $CaCl_2 \cdot 2H_2O$ is the most common version found in brewing. The effect of the addition of this salt can be understood by computing its molecular weight:

Species	Molecular weight (g/mol)
Ca^{2+}	40
$2Cl^-$	70.9
$2H_2O$	36
	146.9

The total of 146.9 g/mol means that 1 mol (*mole*, a scientific unit that describes a set quantity of entities, such as identical molecules) of $CaCl_2 \cdot 2H_2O$ has a mass of (i.e., "weighs" in the common parlance)

Table 1.1 **Carbonate, Bicarbonate, and Carbonic Acid Proportion as a Function of pH**			
pH	**CO_3^{2-}** %	**HCO_3^-** %	**H_2CO_3** %
10	32	68	0
9	5	95	0
8	0	97	3
7	0	81	19
6.5	0	58	42
6	0	30	70
5.5	0	12	88
5	0	4	96

146.9 g. Moreover, each mole of $CaCl_2\cdot2H_2O$ contains 40 g of calcium. Therefore, to increase the calcium content of 1 L of water by 50 mg, it is necessary to add

$$[(146.9 \text{ g/mol})/(40 \text{ g/mol})] \times (50 \text{ mg}) = 183.6 \text{ mg of calcium chloride,}$$

which amounts to 0.025 oz./gal. Conversely, adding 100 mg of calcium chloride to 1 L of water gives an increase of

$$[(40 \text{ g/mol})/(146.9 \text{ g/mol})] \times (100 \text{ mg}) = 27.2 \text{ mg of calcium.}$$

Similar calculations can be used for gypsum ($CaSO_4\cdot2H_2O$), whose molecular weight is 172. Thus,

$$[(172 \text{ g/mol})/(40 \text{ g/mol})] \times (50 \text{ mg}) = 215 \text{ mg of gypsum}$$

is needed to increase the calcium level of 1 L of water by 50 mg, and 100 mg of gypsum added to 1 L of water increases the calcium level by

$$[(40 \text{ g/mol})/(172 \text{ g/mol})] \times (100 \text{ mg}) = 23.2 \text{ mg of calcium.}$$

Calcium carbonate ($CaCO_3$) represents a special case in that not only is it added to increase calcium levels, but also the carbonate ions are used to counteract the acidity in mashes that have significant amounts of dark grains. The relevant molecular weights are

Species	Molecular weight
Ca^{2+}	40
CO_3^{2-}	60
	100

Note that adding 100 mg of calcium carbonate to 1 L of water will potentially contribute

$$[(40 \text{ g/mol})/(100 \text{ g/mol})] \times (100 \text{ mg}) = 40 \text{ mg of calcium ions.}$$

Also, adding 100 mg of $CaCO_3$ to 1 L of water will potentially contribute

$$[(60 \text{ g/mol})/(100 \text{ g/mol})] \times (100 \text{ mg}) = 60 \text{ mg of carbonate ions.}$$

The exact amount will depend on pH (see reactions 1.8 to 1.10 and Table 1.1).

There are many circumstances in which acidification of the brewing water is advantageous. The most common example is with pale beers. Pale malts typically have a pH near 6.0, which is considerably higher than the desired maximum mash pH of 5.4. Therefore, with even mildly carbonate waters, it may be hard to hit mash pH targets without additional water-treatment measures like acidification.

For this purpose, the carbonate content of the raw water is usually monitored, and this content is typically expressed as an equivalent amount of $CaCO_3$. This new unit is denoted with double angle brackets, e.g., $<<HCO_3^- >>$, to distinguish it from the standard unit $<HCO_3^- >$ expressing concentrations in mg/L. The two units are related by a ratio of the molecular weight of $CaCO_3$ (which is 100 g/mol) to that of the ion (61 g/mol in case of HCO_3^- and 60 g/mol in the case of CO_3^{2-}); i.e.,

$$<<HCO_3^->> = [(100 \text{ g/mol})/(61 \text{ g/mol})] \times <HCO_3^-> = \quad (1.12)$$
$$1.64<HCO_3^->$$

and

$$<<CO_3^{2-}>> = [(100 \text{ g/mol})/(60 \text{ g/mol})] \times <CO_3^{2-}> = \quad (1.13)$$
$$1.67<CO_3^{2-}>$$

Given these relationships, alkalinity—the converse of acidity—can be expressed as

$$\text{Alkalinity} = <<HCO_3^->> + 2<<CO_3^{2-}>> .$$

The factor of two in front of the last term reflects the potential of CO_3^{2-} to react with two hydrogen ions via

$$2H^+ + CO_3^{2-} \leftrightarrow H_2CO_3 .$$

Because of the double charge, CO_3^{2-} is said to have an equivalence of 2.

Most water supplies have a pH value between 7.0 and 9.0. For such water, alkalinity is essentially the bicarbonate concentration expressed as an equivalent amount of $CaCO_3$. Moreover, the traditional rule used by brewers of pale beers is that this concentration be below 25 mg of equivalent $CaCO_3$ per 1 L of water, typically expressed as 25 mg/L as $CaCO_3$ (Owades, 1985). To illustrate this relationship, suppose that the raw water has a pH of 8.0 and the alkalinity is 200 mg/L as $CaCO_3$, i.e., $<<HCO_3^->> = 200$. At pH = 8, the bicarbonate ions consist of 97% of the total carbonates (see Table 1.1). To reduce the bicarbonate ion concentration below 25 mg/L as $CaCO_3$, it is necessary to decrease the HCO_3^- until it is

$$\frac{(25 \text{ mg/L})/[(200 \text{ mg/L})]}{0.97} \times 100\% = 12.1\%$$

of the total carbonates. It follows from Table 1.1 that to achieve this concentration, the water pH should be reduced to at least 5.5.

The old practice of biological acidification has taken on a new life in recent years, not only in Germany, but also among some craft brewers in the United States. Since this procedure is used during the mash, a discussion of it is postponed until the end of this chapter.

An alternative is to use food-grade acids. For example, lactic acid ($HC_3H_5O_3$) reacts with calcium carbonate ($CaCO_3$) by transforming it into calcium lactate ($Ca(C_3H_5O_3)_2$) plus other products. The underlying reaction system is

$$2\ CH_3\text{·}\overset{\overset{\displaystyle OH}{|}}{CH}\text{-}\overset{\overset{\displaystyle O}{\|}}{C}\text{-}OH + CaCO_3 \longrightarrow Ca(O\text{-}\overset{\overset{\displaystyle O}{\|}}{C}\text{-}\overset{\overset{\displaystyle OH}{|}}{CH}\text{·}CH_3)_2 + H_2O + CO_2$$

Lactic acid and calcium carbonate form calcium lactate, water, and carbon dioxide

An analogous reaction is the liberation of hydrogen ions (and, hence, the reduction of pH) via

$$2\ CH_3\text{·}\overset{\overset{\displaystyle OH}{|}}{CH}\text{-}\overset{\overset{\displaystyle O}{\|}}{C}\text{-}OH + HCO_3^- \longrightarrow 2\ {}^-O\text{-}\overset{\overset{\displaystyle O}{\|}}{C}\text{-}\overset{\overset{\displaystyle OH}{|}}{CH}\text{·}CH_3 + H^+ + H_2O + CO_2$$

Mechanism for pH reduction

A similar mechanism holds for phosphoric acid (H_3PO_4), which is also widely used. Finally, strong acids like sulfuric (H_2SO_4) are used by large breweries for economic reasons. Regardless of the acid used, the preferred method is to add it incrementally, measuring the pH at each addition until the desired value is achieved. Table 1.2 gives the approximate amounts that will be needed.

A very effective alternative, particularly for small-volume brewers, is to boil alkaline water. Boiling causes bicarbonate ions to precipitate as calcium carbonate and evolve carbon dioxide (as a gas) via

$$Ca^{2+} + 2HCO_3^- \overset{Heat}{\longleftrightarrow} CaCO_3\downarrow + H_2O + CO_2\uparrow . \qquad (1.14)$$

Table I.2	**Acid Treatment of Water**		
	50% Lactic Acid		
Alkalinity	**pH to 6.5**	**pH to 6.0**	**pH to 5.5**
100	13 g/hL	22 g/hL	27 g/hL
200	31 g/hL	50 g/hL	58 g/hL
300	45 g/hL	73 g/hL	86 g/hL
	85% Phosphoric Acid		
Alkalinity	**pH to 6.5**	**pH to 6.0**	**pH to 5.5**
100	7.8 g/hL	15 g/hL	20 g/hL
200	17 g/hL	32 g/hL	41 g/hL
300	26 g/hL	50 g/hL	61 g/hL
	Sulfuric Acid		
Alkalinity	**pH to 6.5**	**pH to 6.0**	**pH to 5.5**
100	4.8 g/hL	7.4 g/hL	9 g/hL
200	10 g/hL	16 g/hL	19 g/hL
300	14.4 g/hL	24 g/hL	28 g/hL

NOTE: From Owades (1985). The unit hectoliter (hL) equals one hundred liters, which is about 22 imperial gallons or 26.4 U.S. gallons (Rabin and Forget, 1998).

It must be emphasized that this reaction mechanism is quite reversible. Hence, some form of aeration (e.g., stirring) is needed to drive the process to the right side of reaction 1.14.

Finally, the old method of treating alkaline water with lime ($Ca(OH)_2$, calcium hydroxide) is sometimes used. The reaction mechanism is

$$Ca(OH)_2 + Ca^{2+} + 2\,HCO_3^- \leftrightarrow 2CaCO_3 + 2H_2O . \qquad (1.15)$$

Calcium carbonate is precipitated in this process.

Another point to be emphasized is that calcium is removed in all of these methods, and its removal must be taken into account. In addition, none of these procedures fully removes carbonates. As a consequence, it is not uncommon for the mash pH to be higher than desired even with water treatment. This condition is particularly prevalent when treatment is applied to water whose alkalinity is in excess of 300 mg/L as $CaCO_3$.

In such circumstances, the notion of a residual alkalinity can be useful. In effect, it gives an estimate of how much more water treatment is needed to reduce the mash pH by a given amount. According to Moll's work that has already been discussed (Moll, 1994), 3.5 equivalents of Ca^{2+} ions are needed to neutralize 1 equivalent of HCO_3^- ions. This statement is the same as saying that

$$\frac{(3.5 \text{ equivalents of } Ca^{2+})}{(2 \text{ equivalents of } Ca^{2+})/(1 \text{ mol of } Ca^{2+})} = 1.75 \text{ mol of } Ca^{2+}$$

is the amount needed to neutralize 1 mol of HCO_3^-. If the contribution from magnesium is ignored, the residual alkalinity (RA) can be defined as follows:

$$RA = <HCO_3^-> - \frac{(1 \text{ mol of } HCO_3^- \times 61 \text{ g/mol of } HCO_3^-)}{(1.75 \text{ mol of } Ca^{2-} \times 40 \text{ g/mol of } Ca^{2+})]} \times <Ca^{2+}> \qquad (1.16)$$
$$= <HCO_3^-> - 0.87 <Ca^{2+}> .$$

This relationship expresses RA in terms of mg/L as an equivalent amount of bicarbonate. Kolbach's (1960) work uses a different measure, German degrees of water hardness. These are obtained by multiplying the RA by 0.046 °German $(mg \cdot L^{-1})$:

$$\text{German degrees of hardness} = RA \text{ (in mg/L)} \times \frac{0.046 \text{ °German}}{mg/L} .$$

Kolbach showed that a reduction of residual alkalinity by 10 German degrees of hardness will reduce the mash pH by 0.3 pH unit. By using this proportion (i.e., 10 °German/0.3 pH unit), one can calculate that to reduce the pH of a 10 °German mash by 0.1 pH unit, the residual alkalinity RA must be decreased by

$$\frac{10\,°\text{German}}{[0.046\,°\text{German}/(\text{mg·L–1})] \times [(0.3\,\text{pH unit})/(0.1\,\text{pH unit})]} = 72.5\,\text{mg/L}. \qquad (1.17)$$

This relationship shows why direct mash acidification is generally preferred to salt additions for the reduction of the mash pH. For example, to reduce the pH of 1 L of mash by 0.1 pH unit requires

$$(72.5\,\text{mg})/0.87 = 83.3\,\text{mg}$$

of additional calcium (see equation 1.16 for the source of the 0.87 value) or, equivalently,

$$(83.3\,\text{mg}) \times [(146.9\,\text{mg/mol})/(40\,\text{mg/mol})] = 305\,\text{mg}$$

of calcium chloride. This approach requires a lot of salt for a rather modest reduction in the mash pH.

In addition to the technical requirements associated with mashing, the salts discussed here are sometimes added to achieve stylistic goals. The guidelines are usually based on the composition of water found at historically important brewing centers. For example, the water associated with Munich and London—both of which have been noted for their dark, moderately hopped beers—is highly alkaline with low concentrations of sulfates, as shown by the following values:

	Munich (mg/L)	London (mg/L)
$<HCO_3^->$	150	160
$<Ca^{2+}>$	75	55
$<SO_4^{2-}>$	10	30

These data reflect three aspects of water-borne minerals that are true in general. First, there is a positive synergism between carbonates from brewing liquor and dark malts. The latter contain complex Maillard products, some of which contribute a rather harsh and biting acidity. The carbonates tend to moderate this characteristic by giving a mellow and "fine malt" palate. Second, hop constituents tend to have the reverse effect. It is easy to demonstrate (Fix and Fix, 1997) that highly hopped beers made with highly alkaline water have a biting and crude bitterness. It is interesting, in this regard, that the London and Munich dark beers have had traditionally low hop levels. Finally, high sulfate levels and dark beers are not a particularly good marriage either. The effects are a drying and astringent afterfinish. From London's and Munich's popular dark beers, one could therefore predict that the sulfate levels of the water associated with London and Munich are low.

One must be careful, on the other hand, with indiscriminate use of historical models, for they can be misleading. A striking case is Dortmund whose water-ion *concentrations* are

$$<HCO_3^->\qquad 180\ mg/L$$
$$<Ca^{2+}>\qquad 225\ mg/L$$
$$<SO_4^{2-}>\qquad 120\ mg/L$$

The Dortmund water is very hard in the sense that it is loaded with minerals. According to M. Jackson (1997), however, Dortmund beers are generally "big and malty, with a clean, delicate sweetness." These characteristics and the Dortmund water-ion concentrations are in direct conflict. Even the most cursory test brew will show that pale lagers made with water having these ion concentrations will have a mineral or salty taste with an underlying harsh aftertaste, totally uncharacteristic of authentic styles. This conflict is resolved in part by noting that Dortmund brewers have always been in the forefront of development of techniques for treating water. Even today, some of the best references on biological acidification come from there. In addition, data from Piendl's (1970–1990) *"Biere aus aller welt"* columns tend to support the notion that extensive mineral reduction is used. For example, the mineral content of Dortmunder Actien–Export has a calcium ion content of 14 mg/L. This concentration is much lower than Piendl's columns mention for just about all the Czech Pilseners, which range from 40 to 75 mg/L. There is of course no direct correlation between finished-beer mineral levels and the mineral content of brewing liquor. Yet, these data tend to support the notion that authentic Dortmunders are brewed with water that is as soft or softer than that used for Czech Pilseners. In short, using the Dortmund water-ion concentrations as a guide to desired water composition for this style is at best dubious. There are many other examples that illustrate analogous effects; Viennese style beers are a prominent example. Thus, instead of using historical examples as a guide, the best overall strategy is to first make sure the technical requirements of the mash are met (i.e., a proper pH) and then to adjust the mineral content by using the finished beer's flavors as the guide.

COMPOSITION OF GRAINS AND MALT

The malt and grain constituents most relevant to brewing can be classified as follows: carbohydrates, nitrogen compounds, lipids, phenols, sulfur compounds, and miscellaneous constituents.

CARBOHYDRATES

A carbohydrate is a compound consisting of carbon, hydrogen, and oxygen molecules. As in any organic compound, carbon is the key ingredient. There is a large and diverse array of carbohydrates in barley, malt, and wort. They typically consist of 70%–85% of the weight of barley and malt and 90%–92% of wort solids. The same types of carbohydrates are also found in other grains such as maize or rice. Fortunately, "erector set models" greatly simplify the discussion of carbohydrate and other organic compound structures. Such models allow the representation of complicated carbohydrates, for example, as a few elementary units linked together in definite ways. Malting and mashing are processes that break down these links (in part) to provide elementary sugars for yeast metabolism.

Glucose is the most important building block in brewing. It has the following structure:

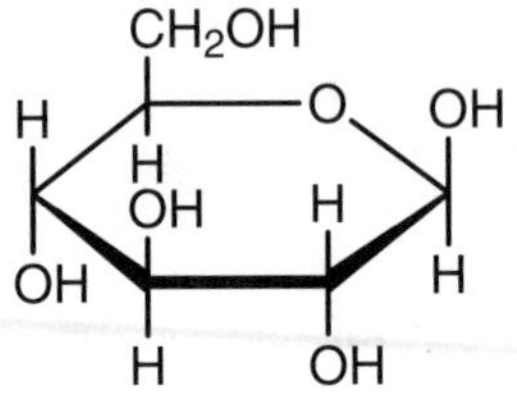

Glucose

One sometimes sees this structure called ß-D-glucose, to distinguish it from different but similar structures. These distinctions are not relevant to brewing and therefore are ignored in this book.

The following widely used shorthand notation omits some of the carbon atoms:

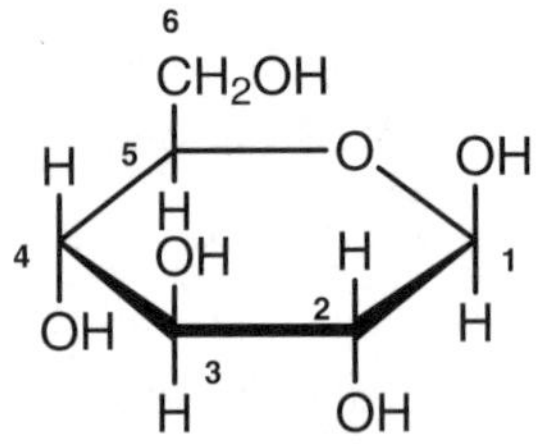

Glucose in shorthand notation

The numbers refer to the carbon atoms not displayed (1 through 5) and the one that is displayed (6). Since this numbering system is universally used, sometimes the numbers themselves are omitted.

One sometimes sees glucose described as a hexose compound, meaning that glucose has six (*hex-*) carbon atoms. In brewing, the greatest significance that glucose plays is its role as a basic building block for other carbohydrates. Because of its basic building block

---

nature, glucose is classified as a *monosaccharide*, with *mono-* referring to a primary unit and *saccharide* referring to a sugar.

Glucose is a minor wort carbohydrate and typically makes up no more than 8%–10% by weight of wort sugars unless it is deliberately added. Of far greater importance are sugars obtained by linking glucose units. Maltose is such an example, and it generally makes up 46% to 50% by weight of the sugars in a grain wort. It has the following structure:

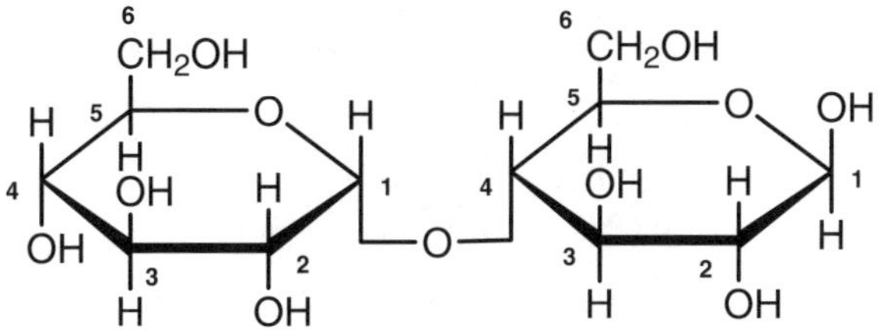

Maltose

Observe the carbon one atom on the left glucose molecule. The second glucose molecule is attached to it via the carbon four atom. Such a combination is given the descriptive name of "1-4 link." A useful shorthand notation for maltose is

$$G-G,$$

where G stands for a glucose unit and the horizontal bar stands for a 1-4 link.

Maltose is called a disaccharide because it consists of two elementary (glucose) units. Maltotriose is a trisaccharide that consists of three glucose units joined by 1-4 links:

$$G-G-G.$$

It is an important wort constituent that typically makes up 12% to 18% of wort sugars. Chains like the following are called amylose sugars and are common in raw grains and malt:

$$G-[G]_n-G.$$

A typical amylose molecule can have more than 1000 glucose units, which are largely separated during malting and mashing. Amylose sugars consisting of four or more glucose units are typically not fermentable by brewing yeasts because of their complexity.

Another important way glucose units can be joined is through a "1-6 link." As the name suggests, two glucose units are linked from the carbon one atom of the first glucose molecule to the carbon six atom on the second glucose molecule in the following way:

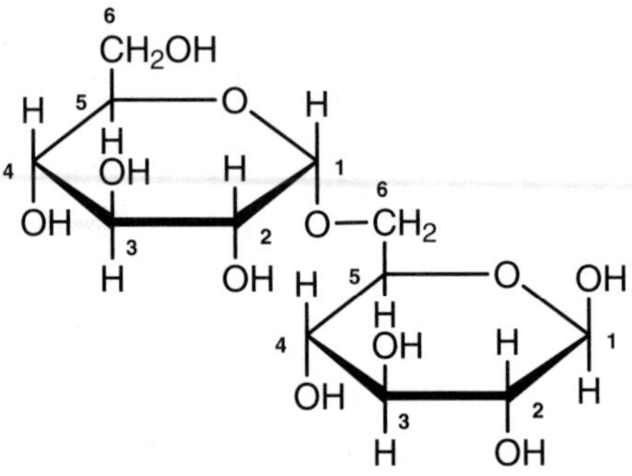

Isomaltose

The shorthand notation for this compound is

$$G$$
$$\uparrow$$
$$G$$

Isomaltose in shorthand notation

The vertical arrow indicates a 1-6 link. Sugars that have both 1-4 and 1-6 links have traditionally been called *dextrins*, although in recent years the term *α-glucans* has become preferred. In biochemical literature, the term *amylopectin* is common. According to Meilgaard (1977), the following are examples of typical α-glucans groups found in wort:

$$G - G$$
$$\uparrow$$
$$G - G - G - G - G$$

Example of group 1 (7 glucous units)

```
                          G — G
                            ↑
          G — G — G — G — G— G— G— G
                    ↑
      G — G — G
```

Example of group 2 (13 glucous units)

```
      G — G — G
            ↑
  G — G — G — G — G— G— G— G— G
                        ↑
          G — G — G— G— G
                      ↑
              G— G
```

Example of group 3 (19 glucous units)

```
              G — G
                ↑
  G — G — G — G — G— G
                    ↑
  G — G — G — G — G— G— G— G— G
        ↑                   ↑
  G — G — G         G — G — G — G — G
```

Example of group 4 (25 glucous units)

Here the bars represent 1-4 links, and the arrows represent 1-6 links. Because of their complexity, these four dextrin groups are not fermentable by normal brewing yeasts, and they are typically passed on to the finished beer. Low-calorie beers are an exception. For these beers, exogenous enzymes are added to the fermenter, and these enzymes break α-glucans down into fermentable sugars. Dextrins in groups 1, 2, and 3 generally will not color with iodine. Those in group 4 will produce a reddish hue when combined with iodine. More complex starch molecules will color a dark blue with iodine.

Carbohydrates more complicated than the four dextrin groups shown are quite common in unmalted grains and in malt. The term *starch* is sometimes used for such molecules. In barley, starch makes up 63%–65% of the dry weight. This proportion is normally reduced to 58%–60% in a properly modified malt. The majority of the "starch conversion" takes place in the mash, and the finished wort will generally have 25%–35% unfermentable carbohydrates.

The second monosaccharide found in wort is fructose. Again, as is customary, the carbon atoms (four in this case) in the "carbon ring" have been left out of the diagram. Unless deliberately added, fructose is a minor wort sugar and typically makes up only 1%–2% of wort carbohydrates.

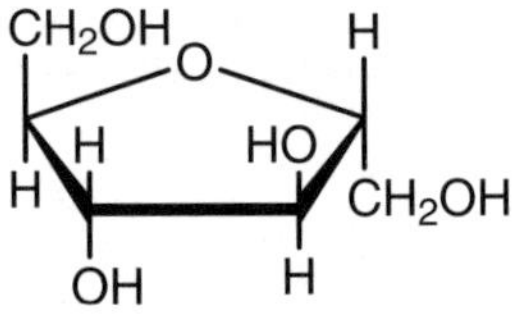

Fructose

Sucrose, common table sugar, consists of a fructose unit joined with a glucose unit. It generally makes up 4%–8% of wort sugars (unless deliberately added) and, in the fermentation, is rapidly "inverted" into glucose and fructose units.

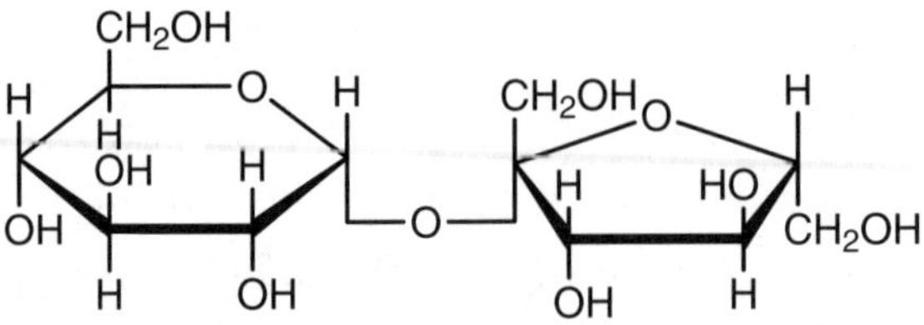

Sucrose

The other group of carbohydrates of interest are gums and cellulose compounds. In brewing, two types—namely, ß-glucans and pentosans—are most important. The ß-glucans consist of glucose units joined with 1-4 links as well as 1-3 links. They have the form

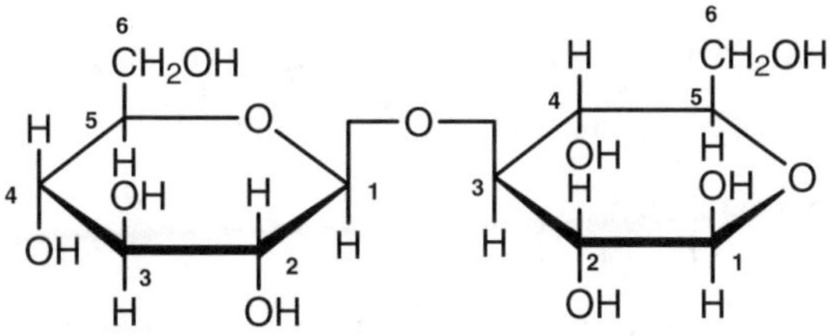

Laminaribiose

LAMINARIBIOSE

Many brewers view ß-glucans as a "wasteful" carbohydrate. They are known to greatly increase wort viscosity when not properly degraded, a situation that can lead to filtration and haze problems (Linemann and Krueger, 1997). For example, they are the major constituent of the sediment of frozen beer (Meilgaard, 1977).

The ß-glucans compounds are found in barley cell walls and can be enzymatically degraded with proper malting and mashing procedures. Their presence in raw barley is one of the main reasons that special enzymes are used when this material is added in appreciable amounts as an adjunct (Bourne et al., 1982). Rye and wheat also have high ß-glucans levels, and these tend to be present even when these grains are malted. A point that is sometimes overlooked is that some of the undegraded ß-glucans can react with iodine. This fact means that an iodine test might show a color reaction even when a satisfactory degradation of malt sugars has been achieved. This effect can be minimized to some extent by straining out grain kernels before adding iodine drops to the sample.

The other carbohydrates of interest are the sugars called pentosans that have five-carbon skeletons. These are formed through linkages of arabinose and xylose:

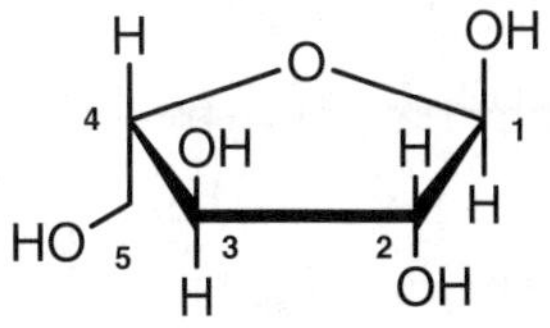

Arabinose

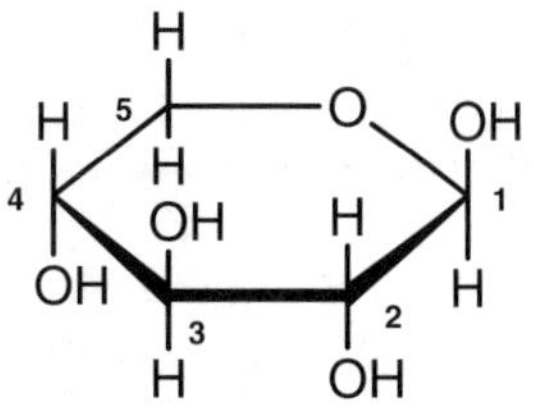

Xylose

Once again, the carbon atoms in the ring (four in arabinose and five in xylose) are customarily omitted from diagrams. As far as brewing is concerned, the pentosans and ß-glucans have similar properties.

It is important to note that ß-glucans levels differ significantly in barley varieties. In general, the cheaper six-row barleys will have a significantly higher ß-glucans content than the two-row varieties (Linemann and Krueger, 1997). Modern brewers have started to specify maximum ß-glucans levels as a part of their malt evaluation. In addition, research in genetic engineering is underway to produce barley with little or no ß-glucans content (Thomas, 1986).

Carbohydrates may also be classified as either reducing or nonreducing (Conn and Stumpf, 1976). Reducing sugars, which are the more common, contain a free (or potentially free) aldehyde (RCHO-type structures) or ketone ($RCOCH_2OH$-type structures) in the molecule. One consequence of this type of structure is the ability to reduce copper salts, and most methods for the estimation of these sugars are based on this property. A typical reaction is

$$RCHO + 2\,Cu^{2+} + 4\,OH^- \rightarrow RCOOH + Cu_2O + 2\,H_2O\,.$$

As far as brewing is concerned, attention has been focused on the reducing sugars present at the fermentation endpoint. De Clerck (1957) estimated that reducing sugars (as an equivalent amount of maltose) consist of 20%–30% of the extract in fermented beer. In addition, he found that the reducing sugars were primarily dextrins, trisaccharides, and pentosans. He concluded that the estimation of reducing sugars in beer was "not of such great importance in the determination of the attenuation limit" (De Clerck, 1957, 2:444–5).

Nevertheless, an assay of reducing sugars is useful in quality control on a comparative basis. For example, for a given beer formulation, there should not be significant batch to batch changes in the reducing sugars at the fermentation endpoint. Variations point to changes, often undesirable, in wort production or with the yeasts being used. The latter effect will be discussed in more detail in chapter 3.

As far as brewing is concerned, the most important sources of nitrogen are amino acids and grain proteins, which in turn are a union of a number of different amino acids (Pierce, 1982). Like carbohydrates, the proteins are degraded into simpler units during malting and mashing.

There are, on the other hand, important differences between proteins and carbohydrates. Although most of the malt starch is converted to simple sugars in the mash, it is during malting that the most protein modification takes place. Also, although approximately the same types of carbohydrates are derived from a wide variety of grains, malted or unmalted, with proteins it is malt that makes the crucial contribution. Unmalted adjuncts such as rice or corn for all practical purposes do not contain proteins that are soluble in wort. Thus, their use will directly dilute the protein content of wort.

The basic building blocks of proteins are amino acids, compounds that have the following structure:

$$\begin{array}{c} H \\ | \\ R-C-COOH \\ | \\ NH_2 \end{array}$$

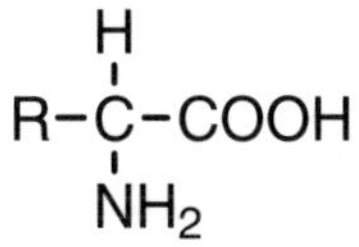

Amino acid

Here R denotes a side chain that will vary with different amino acids. For example, valine has the following structure:

$$\begin{array}{c} CH_3 \quad H \\ | \qquad | \\ H-C-\!\!-\!\!-C-COOH \\ | \qquad | \\ CH_3 \quad NH_2 \end{array}$$

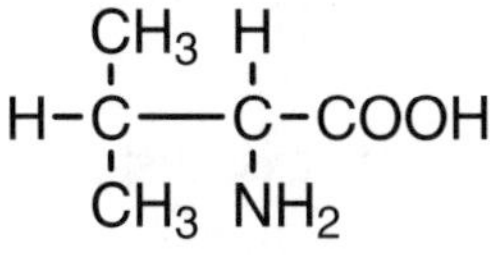

Valine

The importance of amino acids goes well beyond their role as elementary building blocks. For example, they play a crucial role in various Maillard reactions in the malt drying and kettle boil processes

---

(see chapter 2). However, it is during the fermentation that their presence is most significant. As will be seen in chapter 3, yeasts will remove the amino nitrogen component NH_3 (ammonia) via the following diagrammatically shown reaction:

$$R-\overset{\underset{|}{H}}{\underset{\underset{NH_2}{|}}{C}}-COOH + H_2O \longrightarrow NH_3 + R-\overset{\underset{|}{H}}{\underset{\underset{H}{|}}{C}}-OH + CO_2$$

Amino acids break down to ammonia and a carbon skeleton

The amino nitrogen is an essential component of yeast nutrition, and without it, a disordered fermentation results. The carbon skeletons play an important role later in fermentation, particularly in the way by-products like diacetyl and fusel alcohols are formed or repressed (see chapter 3).

Amino acids can be classified in two different ways. One classification is according to the rate they are assimilated by yeasts in a normal fermentation. The classification yields the following four groups:

- Group A. Amino acids that are absorbed very early and rapidly. They are generally eliminated from wort by the end of the growth period.
- Group B. Amino acids that are absorbed at a slower rate.
- Group C. Amino acids that are absorbed only after a considerable lag and only after all of the group A amino acids are utilized.
- Group D. Amino acids that are largely unabsorbed.

In a second classification, it is customary to categorize amino acids according to their roles in the fermentation, above and beyond their contribution of amino nitrogen to yeast nutrition.

- Class 1. Certain amino acids are unimportant because their carbon skeletons can be created by yeasts during normal metabolic activity. It is not necessary for these amino acids to be present in wort.

- Class 2. Other amino acids are vital. Yeasts can in part find replacements, but class 2 amino acids are of such importance that wort must contain them to ensure an adequate supply during the fermentation. Thus, their removal from wort would adversely alter beer flavors.
- Class 3. The last class of amino acids is crucial. Beer wort is the only source of these amino acids, and they are needed in the fermentation. Thus, their removal from wort would significantly alter the flavor of the finished beer.

There are some minor disagreements in the literature concerning the placement of various amino acids in these groupings. Moreover, there are differences in the uptake of amino acids by brewing yeast strains. This point has great practical relevance and is discussed in chapter 3. However, as a general guide, the assignments in Table 1.3 can be used.

Table 1.3	**Amino Acids in Brewing**			
	Absorption rate			
	Group A: Rapid	**Group B: Moderate**	**Group C: Slow**	**Group D: Largely unabsorbed**
Class 1: Unimportant amino acids	Glutamic acid			Proline
	Glutamine			
	Aspartic acid			
	Asparagine			
	Serine			
	Threonine			
Class 2: Vital amino acids		Valine	Glycine	
		Isoleucine	Phenylalanine	
			Tyrosine	
			Alanine	
Class 3: Crucial amino acids	Lysine	Leucine	Tryptophan	
	Arginine	Histidine		

Fortunately, worts that are prepared with reasonable percentages of malt tend to be rich in all amino acid groups. As a consequence, it has become customary to use a single number to characterize the total amino acid content of wort, analogous to the role that specific gravity serves as a measure of the carbohydrate content of wort. For amino acids, the relevant parameter is *free amino nitrogen* (FAN). This is a measure of the nitrogen contributed by the amino acids in the wort, irrespective of type, and is expressed in terms of mg/L.

There is universal agreement that low FAN levels are undesirable in wort (see chapter 3). However, there is some disagreement concerning the minimum desired level. The traditional rule is that serious problems (e.g., long lags in the start of fermentation, high diacetyl levels, etc.) can result if the FAN level goes below 150–175 mg/L (De Clerck, 1957). Differences in yeast strains as well as the presence or absence of other wort nutrients are confounding factors. Nevertheless, a wort in which malt contributes at least 12 g per 100 g of wort (i.e., 12 °P [degrees Plato]) to the carbohydrate pool will typically have FAN levels from 225 to 275 mg/L, which is ideal.

Too much of a good thing is rarely good in brewing, and this statement applies to wort FAN levels. For example, Pogh et al. (1997) reported a decrease in diacetyl as FAN levels increased until a minimum value was achieved. As FAN levels increased further, there was a linear increase in diacetyl with FAN (see also chapter 3). Fusel alcohols are produced from the carbon skeletons of amino acids, and the effect of fusel alcohols on finished beer flavors is quite negative if present above or near their flavor thresholds (see chapter 3). As a general rule, it is usually desirable to keep FAN levels below 350 mg per 1 L of wort, something that can be achieved with a suitable mashing program (Fix and Fix, 1997).

The first class of proteins in order of complexity is the peptides, which are combinations of 2 to 30 amino acids bound by peptide links. They have the following form:

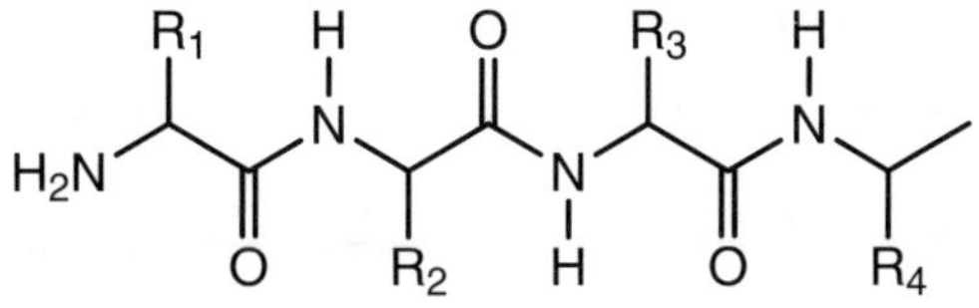

Peptides

That is, CH units are separated by the peptide link:

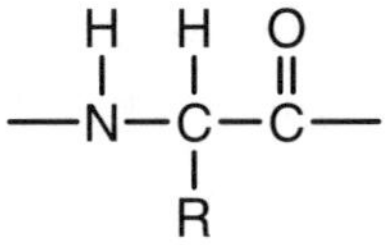

Peptide link

This pattern continues, and polypeptides with thousands of peptide linkages can occur.

It is customary to characterize the amount of protein in barley, malt, or wort either as "% protein" or "% nitrogen." Although there are sources of nitrogen other than protein, the amounts they contribute are very small. Thus % protein is directly proportional to % nitrogen, and the following formula is used:

$$(\% \text{ protein}) = 6.25 \times (\% \text{ nitrogen}).$$

The factor 6.25 is based on the fact that the nitrogen content of protein averages 16%. Thus, multiplying % nitrogen by $100/16 = 6.25$ gives a reasonably accurate approximation of the amount of protein.

Protein levels in different varieties of barley grown in different parts of the world vary widely. They typically range from 8% to 16% protein (1.3% to 2.6% nitrogen). The classic rule, developed by De Clerck (1957), is that barley used for malting should be in the range of 9% to 11% protein (1.4% to 1.8% nitrogen). Although this rule is not always followed, especially in North America, where malt protein levels can approach 13.5% (2.2% nitrogen), it remains a good guideline because excessive nitrogen levels can create a variety of brewing problems. Even in North America, use of high-nitrogen barley is invariably accompanied by the use of significant levels of unmalted grains and syrups that, in effect, dilute nitrogen levels.

During malting, the primary activity is protein breakdown. There is some utilization of amino acids and other loss of protein through precipitation, but for the most part, the protein content of malt is similar to that of the barley from which the malt was made. There is some continuation of protein breakdown in the mash, but the extent is greatly influenced by the mashing schedule used (Fix and Fix, 1997). In addition, there is precipitation of the higher-molecular-weight proteins. Thus wort nitrogen levels are generally lower than those of malt. In addition to amino acids, yeasts can also utilize some proteins, but these are only the simplest peptides having a few amino acid units. Some proteins are precipitated during wort chilling (the so-called "cold break") as well as during fermentation and aging, but most are passed unaltered to the finished beer. The middle-molecular-weight proteins tend to make positive contributions to beer foam, as well as increasing the "maltiness" of beer in ways that are still not completely understood.

The last class of malt proteins, and perhaps the most important, is enzymes. In fact, brewing itself can be regarded as a series of enzyme-mediated reactions (where structures get broken into smaller units), interspersed with some synthesis (the reverse process) primarily mediated by yeasts. Enzymes are relatively high-molecular-weight proteins consisting of about 300–400 amino acids connected by the peptide links already described. Only about 20 different amino acids generally appear in enzymes relevant to brewing (Kunze, 1996). Enzymes are very sensitive to environmental factors—particularly temperature and pH—and tend to be very specific to the substrate on which they act. As a consequence, their names are typically derived from the substance they affect, with an -ase added at the end. For example, ß-glucanase is the enzyme responsible for the degradation of ß-glucans. Because of this specificity, enzymes are best discussed in the context of the reactions they promote.

LIPIDS

In brewing, the most important lipids are fatty acids. They come from three sources. One group is derived from yeast metabolism and is typically saturated, i.e., there are no double bonds like

$$....-C=C-....$$

Example of double bonds

Yeast metabolism and lipid formation are discussed in chapter 3. Oxidized hops are another source of unsaturated fatty acids; hops are discussed in chapter 2.

The group of interest here is the long-chain, unsaturated fatty acids that are derived from malt. They are typically found in wort trub (i.e., particles suspended in the wort), which can consist of as much as 50% lipids (Meilgaard, 1977). Cloudy wort can contain anywhere from 5 to 40 times the unsaturated fatty-acid content of clear wort, an important fact because unsaturated fatty acids can have a significant negative effect even at low concentrations.

On the positive side, fatty acids contribute to yeast viability via a number of mechanisms (see chapter 3), and they also inhibit the formation of some less pleasant acetate esters during fermentation (see chapter 3). On the negative side, they work against beer foam stability, as any fatty material does. Even more significantly, they play an important role in beer staling (see chapter 4).

A number of fatty acids are derived from malt, but the most important are oleic and linoleic acids. These have 18 carbon atoms and one and two double bonds, respectively (Conn and Stumpf, 1976):

$$\underset{CH_3(CH_2)_6CH_2}{\overset{H}{\diagup}} = \underset{CH_2(CH_2)_6CO_2H}{\overset{H}{\diagdown}}$$

Oleic acid

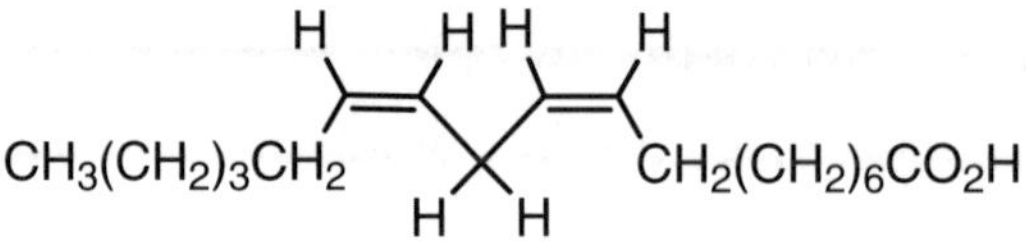

CH3(CH2)3CH2 ... CH2(CH2)6CO2H

Linoleic acid

These are distinguished from aldehydes and alcohols by the characteristic COOH tail. It is important to point out here that their presence in wort is at the center of the controversy concerning wort clarity. Unsaturated fatty acids have both positive and negative effects. Thus, some investigations have reported that wort clarity (via trub removal) is essential (Zangrando, 1979), whereas other investigators have found some carryover of unsaturated fatty acids in the trub to be beneficial (Hough et al., 1981). A traditional practice in North America involves using starting tanks to strike a compromise between these effects. The partially clarified wort is combined with yeasts in the starter tanks and left for 12 to 24 hours. It is then transferred into fermenters, leaving most of the trub behind in the cold break (Knudsen, 1977).

PHENOLS

Phenols have ring structures; the following diagram shows a typical structure:

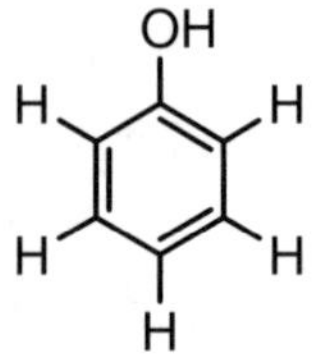

Phenol

Here for simplicity, the carbon atoms have not been displayed. In addition, shorthand approaches like

Aromatic ring

are sometimes used for these compounds.

There are a large number of phenolic compounds, but the simplest ones relevant to brewing have two aromatic rings. Catechin (sometimes called catechol) is an example:

Catechin

More complicated structures are polymers (i.e., repeating structures) of such elementary phenols, exactly as carbohydrates are built from monosaccharides and proteins from amino acids. These complex phenols are called polyphenols in the brewing literature and flavonoids in some other disciplines (Dadie, 1980).

The term *tannin* is often used. Historically, tannins are phenolic materials applied to animal hides for tanning. In the modern brewing literature, the term generally refers to polyphenols that have molecular weights in the range of 500 to 3000. Of special interest are those having 4–25 aromatic rings (Dadie, 1980) since these can bond with high-molecular-weight proteins and form haze.

In addition to beer haze, phenolic compounds play a major role in other areas affecting the quality of the finished beer. The role of phenolic compounds in beer oxidation is of great interest. It has been shown that they can act as both oxidizing and reducing agents, depending on circumstances (discussed in detail in chapter 4). In addition, it is known that phenolic compounds in their reduced state tend to add an element of freshness to beer (Meilgaard, 1977), and this is considered an attractive part of the flavor of young beer. However, when oxidized, many phenolics contribute very astringent and harsh tones. The term *herbstoffe* has traditionally been used to describe the grain bitterness found in some beers, as contrasted with the more mellow bitterness that hop α-acids contribute.

Phenolic compounds derived from malt are typically found in the outer layers of the kernel, with the husk making notable contributions. Since two-row barley has a thinner husk than six-row barley, the malt these barleys produce contains different amounts of phenolics.

Extraction of phenolic compounds occurs to some extent during mashing, but it is in sparging that the issue becomes critical (Prechtl, 1964; Schneider, 1997). This is controlled primarily by the following factors: volume of sparge water used, the instantaneous pH of the runoff collected from the sparge, the alkalinity of the sparge water used, and the temperature of the sparge water.

Because of simple mechanical extraction, it is clear that phenols will increase with the volume of sparge water used. A parameter that correlates more directly with the grainy bitterness in the finished beer is the pH of the runoff collected. The first increases in direct proportion to the second. This relationship is why highly alkaline sparge waters must be avoided, unless there is balancing acidity from dark malts in the grist.

It should be pointed out that some phenolic compounds found in beer are derived from sources other than grains. Contributions from hops are discussed in chapter 2. Residues from halogen-based (e.g., chlorine or iodine) sanitizing agents if absorbed in beer can lead to very nasty flavored phenolic compounds, whose effect is far out of proportion to their concentration. Arguably, the most important contributor to phenolic compounds in finished beer comes from yeast metabolism. This very important topic is discussed in chapter 3.

SULFUR COMPOUNDS

The most important sulfur compound is dimethyl sulfide (DMS). It has the following structure:

$$H_3C-S-CH_3$$

DMS

As is typical of sulfur compounds, DMS has very powerful flavors. An appropriate flavor threshold is approximately 0.03 mg/L, and because of this low threshold, DMS levels are normally reported in μg/L (i.e., parts per billion; μ means *micro-* or *one-millionth*). In each liter of water there are 1000 g of water (a close approximation at room temperature and pressure), so 1 μg/L = 1 μg/L $\times$ [(1 L)/(1000 g)] $\times$ [(1 g)/(1,000,000 μg)] = 1/1,000,000,000 or 1 part per 1 billion parts). Note that 0.03 mg/L = 30 μg/L.

The most important DMS precursor is S-methyl methionine (SMM); it is formed in malt during germination. SMM has the following structure:

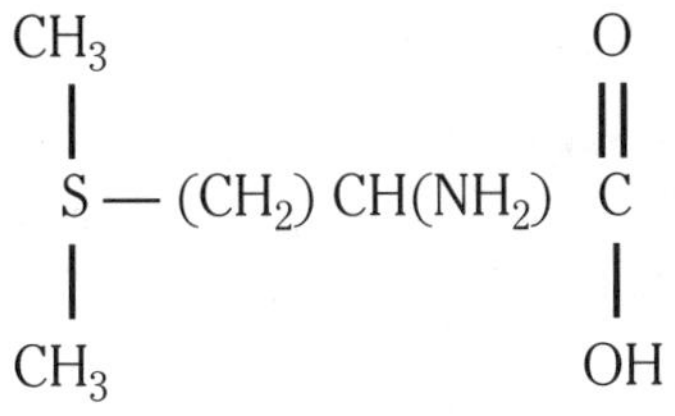

$$
\begin{array}{ccc}
CH_3 & & O \\
| & & \| \\
S-(CH_2)\ CH(NH_2) & & C \\
| & & | \\
CH_3 & & OH
\end{array}
$$

SMM

SMM is broken down to DMS via heat, mainly during the kettle boil and wort cooling period (see chapter 2).

Different types of malt have dramatically different SMM levels and thus produce dramatically different DMS levels in the finished beer. For example, ale malt has minimal SMM levels because this DMS precursor is largely removed by the high kilning temperatures used to produce ale malt (Fix, 1992). As a result, British ales brewed from these malts typically have DMS levels in the range of 0 to 20 μg/L, i.e., at most a very dilute secondary constituent (Fix, 1992). The same is typically true of ales brewed in North America, even those using domestic malt with higher SMM values than their British counterparts. This result is due to the strong volatilization of DMS that occurs in ambient-temperature ale fermentations.

Lager malt, in contrast, is produced with lower kilning temperatures and as a consequence has much higher levels of SMM. The extent to which this is converted to DMS in the finished beer depends on brewing conditions (see chapter 2). The cold fermentation and maturation also encourage the stability of DMS. Traditionally, DMS levels in lager beer have varied from 30 to 90 μg/L, although higher levels have been reported, particularly with German pale lagers and in the products of some of the regional breweries in the United States. During the 1990s, DMS levels in lagers have been considerably lower. Kunze has cited 60 μg/L as the "objectionable threshold" for German lagers (Kunze, 1996). Many large industrial brewers both in the United States and United Kingdom produce lagers with DMS levels well below 30 μg/L in order to obtain a beer with a cleaner finish. The descriptor "cat urine" (introduced by Bamforth, 1998) for DMS above its threshold of 30 μg/L gives some indication of the negative reaction that some have when DMS is detected. Lager aficionados, on the other hand, tend to be critical of low-DMS beers because of their very low malt profiles. A good compromise is for DMS to be in the range of 30–60 μg/L, so that malty or slightly sulfury lager flavors will be present, without the distracting cooked-corn tones that are present with higher DMS levels.

DMS can also be created by bacterial action, the most common being wort-spoiling coliform bacteria (see chapter 3). When bacteria produces DMS, its levels can be much higher than the range of 30–60 μg/L (Owades and Plam, 1988) and thus can impart sulfury flavors reminiscent of overcooked sulfur-bearing vegetables like broccoli or cabbage. It was once believed that DMS could be created in the fermentation by normal brewing yeasts. The precursor in malt is dimethyl sulfoxide (DMSO):

$$H_3C \overset{\overset{\textstyle O}{\|}}{\underset{}{S}} CH_3$$

DMSO

Although DMSO can be reduced to DMS by brewing yeasts in synthetic media (White, 1977), the minor sulfur constituents (e.g., methionine sulfoxide) in beer wort block this pathway for the most part (Gibson and Bamforth, 1982). Therefore, in fermentations free of bacterial infections, DMS levels are actually reduced when this sulfur constituent is scrubbed out of beer by evolving CO_2 gas.

The second most important sulfur constituent is hydrogen sulfide (H_2S). An appropriate threshold has been reported as 10 µg/L in lagers and 30 µg/L in ales (Fix, 1992). It is produced by yeasts during the fermentation, but is rapidly removed by evolving CO_2 gas. Thus, it is rare to find its characteristic rotten-egg flavor tones in beer. Normal H_2S levels are usually reported in the range of 0 to 0.9 µg/L (Fix, 1992). Various bacteria, especially the wort spoilers, can create very high H_2S levels, which are normally accompanied by large DMS levels. The net flavors so produced are highly sulfury and have been variously described as reminiscent of parsnip, celery, cooked corn, cooked cabbage, or black currant. Finally, elevated pitching rates (e.g., initial yeast-cell concentrations above 100 million cells per milliliter) can lead to offensive sulfur compounds that spill over in part to the finished beer and have a negative impact on beer flavor (Nyborg et al., 1999).

Mercaptans or thiols are sulfur compounds having H_2S as a base; they have the following form:

$$R-S-H$$

Mercaptan or thiol structure

Some have very powerful flavors. For example, amyl mercaptan has a threshold near 0.00007 µg/L = 0.00000007 mg/L! Some of these compounds are derived from malt, hops, and additives like bisulfites. The extent to which they are formed in the fermentation is strongly dependent on the yeast strain used (see chapter 3).

MISCELLANEOUS CONSTITUENTS

Barley and malt are rich sources of several vitamins. These include the B-complex vitamins, which are important growth factors for yeasts during fermentation (see chapter 3). In this group, biotin, inositol, and pantothenate are important. Fortunately, worts with a high malt content tend to supply these vitamins in at least a twofold excess of what yeasts need (Meilgaard, 1977). The vitamins left after the fermentation add to the overall nutritional value of beer.

Yeasts also require trace quantities of copper (0.012 ppm), iron (0.075 ppm), and zinc (0.5 ppm), although levels in excess of these can lead to other problems. Most grain worts are sufficient in copper and iron, and all-malt worts tend to be sufficient in all three metals (Meilgaard, 1977). Many of the so-called "yeast nutrients" contain mixtures of these metals. However, some brewers directly add zinc to correct deficiencies that typically arise from high concentrations of unmalted grains.

Magnesium ions also play an important role in yeast growth, primarily as a cofactor in metabolic reactions. Malt generally will provide sufficient magnesium for these purposes, even when the brewing water is low in this ion. Corrections with $MgSO_4$ additions are needed only with very high adjunct worts. It is interesting that calcium ions, which are highly beneficial in the mash, tend to inhibit yeast growth if they are at excessive levels (Saltukoglu and Slaughter, 1983).

There are numerous trace elements derived from grains that are difficult to measure and whose effect on beer flavor is uncertain. Among these are nucleic acids and amines. There are also trace elements that are known to be important. Products of the browning reaction, also called the Maillard reaction, certainly fall into this category. This reaction is important not only in beer, but also in any perishable food. The reaction, described in detail next and further in chapter 2, occurs in brewing during the malt kilning as well as during the kettle boil. Products of this reaction—typically called melanoidins—can be found in malt, particularly dark malts.

TRANSFORMATIONS DURING MALTING AND MASHING
BASICS OF MALTING

In its simplest terms, malting is the controlled germination of cereal grains. The basic goals of this process are the following:

- Enzyme development. The appropriate enzymes are generally not found in unmalted grains and must be formed during malting.

- Cytolysis. The grain cell walls must be broken down so that enzymes can start modification.
- Modification. Modification involves the appropriate degradation of protein and starch so that they can be used advantageously in mashing and in the fermentation.
- Maillard products. These products are responsible for the coloring potential of malt and also for the special flavors that are associated with these products.

Malting starts with steeping so that the grain kernels can absorb water. This is needed for growth of the acrospire into the embryo (see Fig. 1.1). Also, foreign material is washed from the kernels. The moisture level is normally increased to 38%–42%, but can go as high as 46%–47% for problematic barley and as a way to reduce the malting times. High-moisture steeps, however, do not usually produce high-quality malts.

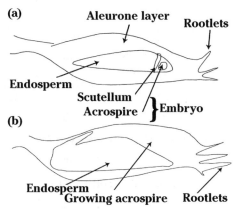

Figure 1.1. (A) Raw barley and (B) germinating (sprouting) barley.

Since the kernel is not yet a green plant, the oxygen needed for growth and respiration is supplied by the maltster during steeping (Sebree, 1997). Respiration consumes part of the barley carbohydrates via the standard respiration pathway:

$$\text{Glucose} \rightarrow \text{Pyruvate} \rightarrow CO_2$$

(See also chapter 3 for a discussion of the basic metabolic pathways.) As carbohydrates in the embryo are depleted, the scutellum tissue begins to produce various hormones (gibberellic acid and related compounds), which in turn stimulate enzyme formation. Exogenous additions of gibberellic acid during malting are sometimes used as a short cut; however, this practice is controversial and banned in Germany and other countries.

The following are the four basic enzyme systems that are crucial to both malting and mashing:

- Starch-degrading enzymes consist of α- and ß-amylase as well as select debranching enzymes (i.e., enzymes capable of breaking 1-6 links connecting glucose in grain carbohydrates).
- Cytolytic enzymes are responsible for degrading cell-wall structures. The most important of these enzymes are ß-glucanase and cytase.
- Protein-degrading enzymes consist mainly of proteinase (which works on a wide spectrum of proteins) and peptidase (which mainly degrades medium-molecular-weight proteins to peptides and amino acids). These proteolytic enzymes are also active in mashing.
- Acid-forming enzymes are needed for control of malt and mash pH. An important example is phosphatase, which breaks down phosphoric acid.

Once steeping is complete, the kernels are removed from the steeping water. Here, germination—which in fact has already started during steeping—is encouraged. Oxygen is still supplied to promote respiration.

Cytolysis is crucial to all aspects of grain modification. The kernel cell walls are essentially composed of ß-glucans and other gums already discussed. They are responsible for the hardness (friability) of the endosperm in unmalted barley, and the cell walls must be degraded by the action of hydrolytic enzymes in order for the carbohydrases and proteolytic enzymes to attack the starch and protein molecules. The extent of cytolytic modification is usually measured in two ways. Friability (as measured by a friabilimeter) is a widely accepted malt-quality index, and the usual rule is for it to be no less than 80%. Brewers are also sensitive to the viscosity of standardized worts, and values in the range of 1.4–1.5 mPa•s are considered desirable.

The second most important aspect of malt modification is the degradation of proteins. Actually, proteins—in addition to gums—are structural components of the endosperm cell walls. Thus, cytolysis and protein modification are not independent. During germination, 35% to 40% of the protein is degraded to soluble compounds (Kunze, 1996). The ratio of soluble protein to total protein—called the Kolbach index in Europe and the S/T ratio in the Americas—gives a reasonable index of the extent of protein degradation. Traditional lager malt used to fall into the range 36%–38%, although today, all malt types are typically over 40%. Values in excess of 45% are generally unacceptable since they imply overdegradation of protein, which tends to have a negative impact on the finished beer head stand and malt flavor. Another interesting index, but unfortunately one that is rarely found on malt-analysis sheets, is the malt FAN level. In modern highly modified malts, the FAN levels can fall into the range 150–250 mg/L, which is more than adequate for wort.

Carbohydrate modification is considerably less extensive. Kunze (1996) reported typical values, which are shown in Table 1.4. From

Table .1.4	**Starch Content of Barley and Malt**	
	Starch (%)	**Sugar (%)**
Barley	63	2
Malt	58	8
NOTE: From Kunze (1996).		

these, it can be seen that barley begins with a starch level of 63%; during malting, some of the barley starch is degraded into simple sugars with the result that the malt's starch content is only 58%. Compared to barley, the malt's sugar content is higher, however.

As far as malting is concerned, the most important carbohydrase enzymes are the debranching types. These are able to break the 1-6 links already discussed.

$$
\begin{array}{l}
\text{G-G-G-G-G-G*} \\
\quad\uparrow \\
\text{G-G-G}
\end{array}
\quad\rightarrow\quad
\text{G-G-G-G-G-G*} \quad + \quad \text{G-G-G*}
$$

Reaction involving debranching enzymes

G* denotes the reducing end of the chain. As a result of the action of the debranching enzymes, the average length of α-glucans (amylopectins) is reduced, and the amount of straight-chain amylase starches is increased.

The second most important enzyme is ß-amylase. The enzyme breaks 1-4 links near the reducing ends and produces maltose units:

$$
\text{G-G-G-G-G-G} \quad\rightarrow\quad \text{G-G-G-G*} \quad + \quad \text{G-G*}
$$

Production of maltose units by ß-amylase

The ß-amylase does not attack 1-6 links nor will it break 1-4 links at reducing ends that are near a 1-6 link:

$$
\begin{array}{l}
\text{G-G-G-G-G-G*} \\
\qquad\uparrow \\
\quad\text{G-G-G}
\end{array}
$$

Blockage of ß-amylase at a 1-6 links

The enzyme α-amylase is generally more important to the mash than in malting. This enzyme will break 1-4 links at random:

$$
\begin{array}{l}
\text{G-G-G-G-G-G-G-G*} \\
\qquad\uparrow \\
\quad\text{G-G-G}
\end{array}
\;\rightarrow\;
\text{G*} + \text{G-G*} +
\begin{array}{l}
\text{G-G-G-G-G*} \\
\qquad\uparrow \\
\quad\text{G-G-G}
\end{array}
$$

Action of ß-amylase

The α-amylase and ß-amylase enzymes are active in mashing, where the bulk of the carbohydrate modification takes place.

A widely used index for starch degradation is the fine- to coarse-grind extract difference. The values normally fall in the range 1.5–2.0. Values in excess of 2.2 are a cause of concern—particularly with respect to the uniformity of grain modification. It must be emphasized that such a situation can be present in malt whose other modification indices are in acceptable ranges. The total extract (fine grind or coarse grind) is usually quoted on malt-analysis sheets. It is usually expressed as yield, i.e., the percentage of the extract recovered as a proportion of the grain's weight. These values are obtained under laboratory conditions, and it is not possible or desirable to achieve them in a full mash.

Malting terminates in the kiln, where moisture is removed by drying. This is a crucial step, since malt with a high moisture level does not store well. In extreme cases, mold and/or unattractive aromas (e.g., those recalling damp, cut grass) can be detected. In less extreme cases, there will still be negative effects on the quality of malt flavor. A widely accepted rule is for the moisture level of malt to be below 4% by weight. Because of its importance, it is almost always quoted on malt-analysis sheets. It should be emphasized that a malt's moisture level is not stable in storage and that the malt will invariably take on the ambient moisture level. This process can undo all of the effects of the kiln and lead to a wide range of deterioration products. It is highly advisable that malt that is stored over an extended period of time have its moisture level checked on a regular basis. Effective techniques for checking the moisture level can be found in *Methods of Analysis* (American Society of Brewing Chemists, 1987).

The color of malt is also determined during the kilning, and this process in effect defines the malt type. Different units to describe malt color are used in Europe and the United States. The European Brewing Congress uses °EBC, and the American Society of Brewing Chemists uses °ASBC. (This unit is sometimes quoted as °SRM,

Standard Reference Method or °Lovibond.) Between these two terms there are differences as well (Fix and Fix, 1997), but for the purpose of this discussion, they will be ignored.) Since these units are obtained from different determinations, they are not strictly comparable. Table 1.5 gives approximate correspondences. The errors are less than 5% in the lower color range of the data in Table 1.5 and increase to 20%–25% in the upper color range.

Malt color is controlled during the kilning by means of the temperature used. There are generally two phases (Kunze, 1996). The first is the drying phase, which is conducted at relatively low temperatures. Next is the curing phase during which the bulk of the color formation is achieved. There is also deactivation of enzyme systems during curing, and the extent of those that survive is inversely proportion to color. Another important factor is that high-protein malts tend to have more enzymes. The strength of a malt's enzyme system is usually called *diastatic power* (DP). The °ASBC procedure described in *Methods of Analysis* (American Society of Brewing Chemists, 1987) is used in the Americas, and the units are called °Lintner (°L). Typical temperatures and color levels for the classical malt types are given in Table 1.6. A different but closely related determination is used in Europe, and the units are called °W-K (W-K = Windisch-Kolbach) (Kunze, 1996). These

Table 1.5	**Approximate Color Correspondences**	
	°ASBC	**°EBC**	
	1.5	2.5	
	2	4	
	3	6.5	
	4	9	
	6	14	
	8	20	
	10	24	
	15	30	

Table I.6	**Typical Malt Data**		
Malt type	**Curing temperature (°C)**	**Color (°ASBC)**	**DP (°L)**
Czech Pilsener	85	1.4–1.8	90–100
Domestic 2-row malt	87.5	1.4–2.0	125–135
Pale-ale/Vienna	90	3.0–4.0	50–70
Munich	100–105	6.0–10.0	40–65
NOTE: From Fix and Fix (1997).			

two units are related to a reasonable degree of accuracy (generally better than 1%) by the relationship $(°W\text{-}K) = 3.5 \times (°L) - 16$.

Not all enzyme destruction is harmful. Lipid-active enzymes (e.g., lipase and lipoxygenase) are major players in beer staling, so their removal is highly advantageous. These issues are discussed in chapter 4.

Important flavor compounds are formed during the kilning, and the strength and character of these increase with curing temperatures and hence color. These compounds are the Maillard products already mentioned; they are formed through a process called *nonenzymatic browning*. There are two steps in this process (Holtermand, 1963):

1. Amadori rearrangement is a combination of a malt sugar and low-molecular-weight proteins (typically an amino acid) to form a protein-sugar complex. This process requires heat and generally will not take place below 100 °C.

2. Strecker degradation is the breakdown of the Amadori complex into aldehydes such as furfural (isobutyraldehyde and isovaleraldehyde are also important). Heterocyclic compounds such as simple melanoidins are also formed.

These have a toast-like or malty aroma and a reddish-brown color; they taste acidic.

Because of the temperature threshold of (approximately) 100 °C for nonenzymatic browning, these products are more relevant to Munich malts than to pale malts. However, the Maillard products can also be created during wort boiling, depending on the amount of thermal loading applied. Here (and elsewhere in this book), *thermal loading* refers to the area under the temperature vs. time curve that describes a given process. In malting, this term refers to the times and temperatures used during the kilning process. In chapter 2, analogous considerations arise with wort boiling (Herrmann et al., 1985).

Roasted malts have the largest concentrations of Maillard products because of the elevated temperatures used (Blenkinsop, 1991). In addition to simple melanoidins, higher heterocyclics are formed. Two examples are pyrroles and pyrazines. They have the following structure.

Structural pyrazine (left) and pyrrole (right)

These compounds usually take on burnt or sharp, acidic flavor tones. On the other hand, the lower heterocyclics, like furfural, recall caramel and toffee flavor tones. A survey of the roasted malts and their properties can be found in the reference by Fix and Fix (1997).

Sulfur compounds and their precursors (denoted DMS-P) are formed during malting and have a profound impact on finished beer flavor as already noted. The important DMS-P (SMM and DMSO) are formed during germination, and it is now established that the extent is proportional to modification (White, 1977). Barley variety is also an issue; for example, high-protein six-row barley tends to have elevated levels (Anners and Bamforth, 1982). The final factor is malt kilning during which DMS-P is converted to DMS. The latter is highly volatile and, by and large, is removed soon after it is formed. Therefore, DMS-P levels are not usually an important issue for highly modified malt if it is kilned at high temperature (e.g., pale ale, Vienna, Munich). It is, however, very much an issue for highly modified Czech Pilsener malts as well as high-protein pale malts. These issues are discussed in chapter 2 as there is additional DMS reduction during wort boiling.

So far, this discussion has focused on malting one grain, namely, barley. However, other grains can and have been malted for brewing. Most notable is wheat, but rye and sorghum are also important for special beer types. The first two grains (rye and wheat) have common characteristics including high protein levels and very high ß-glucans levels. Cytolytic activity is crucial, and both malt types challenge the skill of malting companies. Appropriate mashing procedures are also important. Malted sorghum is popular in Africa for that continent's indigenous beer styles. These are beers are sour because of the way they are fermented. Yet, the malting issues associated with these grains are quite interesting, and important research projects have been devoted to this topic (Diefenbach, 1996).

Mashing is best seen as an extension of malting because at each step in the mashing process, decisions about the way to proceed cannot be made in isolation from the ways the grains were malted. The starting point is when the grains are milled. Ideally, individual husks are cracked open, exposing the starchy endosperm yet leaving the husk intact (Fix, 1994). The latter is needed as a filter medium during lautering (the process of straining out the spent grains from the sweet wort; Rabin and Forget, 1998). Moreover, husks contain astringent materials, which can be extracted during sparging. This effect is not subtle and can be demonstrated in test brews (Prechtl, 1964; Fix, 1992). Since quality malts have only a small difference in fine- and coarse-grind extract—less than 2%, as already noted—small-scale brewers tend to prefer coarser grinds, since the difference in the extract with a finer crush is small enough to be justified. Large-volume brewers, on the other hand, use hundreds of thousands of tons of malt each year, and in this context, a 2% swing in extract recovery is a major issue. Therefore, extremely fine grinds using hammer mills are finding favor (Kunze, 1996). In place of the traditional lauter tun (the specially designed container for straining out the spent grains), new technology—most notably the Mash Filter 2001™ (Rehberger and Luther, 1994)—is used. The combination is not only capable of producing normal wort with high extract efficiency, it also permits 10–12 brews per day.

Many mashing systems are used throughout the world. Given the high degree of modification of modern malts, brewers have a lot of options. The following key parameters can be controlled: time and temperature, pH, and mash thickness.

The effects of holding the mash in different temperature ranges—called *rests*—are summarized in Table 1.7. The lowest temperature range (35–40 °C) allows for extensive ß-glucans breakdown, the effect of this being improved lautering and beer clarification. Between 60 and 65 °C, α-amylase acts as a liquefaction enzyme and encourages the dispersal of grain carbohydrates, proteins, and enzymes into the mash. This activity greatly assists enzymatic activity at the higher temperatures; hence, rests in this range invariably give higher yields. Generally, 15 min is adequate to achieve all of these effects, and certainly, no more than 30 min is required (Fix and Fix, 1997). A rest at 35–40 °C was once called the "acid rest" since phosphatases and other acid-forming enzymes are active in this temperature range. However, practical experience has shown that the mash pH is quickly established and then is held in check from that point forward by strong buffering agents in the mash. Thus, other measures are needed to establish and control mash pH.

The proteolytic regime starts near 45 °C and continues up to 55 °C, with the strongest activity being near 50 °C. There is some activity outside this range, but it is small and not of practical significance.

Table 1.7	Mashing Effects during Rests at Various Temperatures	
Temperature (°C)	Effects promoted during mash rest in given temperature range
35–40	Grain liquefaction and ß-glucanase activity
45–55	ß-glucanase activity
47–52	Proteinase and peptidase activity
55–60	Termination of ß-glucanase activity with minimal proteinase and peptidase activity
60–65	α-amylase activity leading to maltose production
65–70	ß-amylase activity leading to breakdown of starch to dextrins
70–72	Formation of glycoproteins leading to stability and texture qualities of beer foam

With malts having only a relatively small amount of protein degradation (e.g., traditional lager malts), a rest in this range is highly advantageous, particularly if unmalted cereals are used because these in effect dilute the FAN pool. Traditional lager malts are rare in modern times, and if the Kolbach index is 40% or more, rests in the proteolytic range can create negative effects, including poor head retention and/or dull malt flavors. As a consequence, in Germany, high-temperature, short-time mashing schemes (called *Hochkurz Maischverfahren*) are being used by many commercial brewers. These schemes usually have rests near 60 °C and 70 °C with a total mashing time under one hour.

In general, the rests above 60 °C are by a wide margin the most important to mashing (Lewis, 1998). This is the regime in which the bulk of the extract is dissolved and the percent fermentability is determined. This process begins with gelatinization, in which carbohydrate granules are literally burst open, making them more accessible to enzymatic attack. This process starts at low temperatures, but goes to completion when temperatures of at least 60–65 °C are achieved. As a consequence, it will occur naturally with malted barley in just about any reasonable mash. In contrast, unmalted cereals require boiling or hot flaking to achieve proper gelatinization.

Temperature plays a crucial role in the type of carbohydrate profile achieved (see Table 1.8). For example, mashing at 60 °C will yield about 80% fermentable sugars. For beers of normal strength, the apparent attenuation (i.e., the degree of fermentation as measured by a hydrometer without alcohol corrections) is approximately a factor of 1.2 higher than the real attenuation (i.e., the actual amount of sugars fermented). For example, 80% real attenuation is equivalent to 96% apparent attenuation. German diat Pils are mashed at 60 °C (see data in Piendl, 1970–1990). At the other extreme, mashing at 70 °C will yield a real attenuation (actual fermentable sugar percentage) of 60%, which corresponds (approximately) to an apparent attenuation of $(0.6 \times 1.2 \times 100\%) = 72\%$. The difference in maltose produced follows the same pattern.

Another point that is a relatively new discovery concerns glycoproteins. These are polymers of dextrins and middle- to high-molecular-weight proteins. They are known to be major factors in the stability of beer foam (Melm et al., 1995). Moreover, the complexes are formed when the mash temperature falls into the range 70–74 °C with 72 °C being preferred (Ishibashi et al., 1997). Test brews have shown

Table I.8	**Sugar Profiles**		
Percentages of Sugar Compounds at Various Temperatures			
	60 °C	**65 °C**	**70 °C**
Disaccharide	61	55	41
Trisaccharide	9	12	16
Monosaccharide	10	9	8
Dextrins	20	24	35
NOTE: Adapted from Fix and Fix (1997).			

that this is a significant effect. Mash thickness is the concentration of grains in the mash. It is normally expressed as a liquor:grist ratio; i.e., thickness = liters of water per kilogram of grains. Many brewers quote thickness in terms of pounds per gallon.

Mashes as thin as 3.5 L/kg (2.4 lb./gal.; note that the SI [metric] units are given in terms of volume per mass whereas the U.S. customary units are given in terms of weight per volume) have been reported as well as mashes as thick as 1.5 L/kg (5.5 lb./gal.) (Lewis and Young, 1995). Normally, mash thickness is near 2.67 L/kg (3.1 lb./gal.). It has been noted that thicker mashes favor proteolytic activity, whereas thinner mashes favor carbohydrase action because of the restraining influence sugar concentrations have on α- and ß-amylase (De Clerck, 1957). Measured data (Lewis and Young, 1995) show, however, that these effects are small over the 2–3 L/kg range.

The final control parameter in the mash is the pH, which has been known for a long time to be very important (De Clerck, 1957). The classic rule is for the chilled wort to have a pH of 5.0 to 5.2 and, to achieve this level, it is desirable to establish a mash pH in the range of 5.2–5.4 (Hind, 1950). This range, first of all, is favorable to enzymatic activity. Although the various enzymes have different optimum pH levels, in practical brewing conditions the enzymes' activities do not decrease by much if the pH levels are more acid On the other hand, there is typically a sharp decrease in the enzymes' activity if the pH becomes more basic. This factor is illustrated for enzymes, by using amylase as an example, in Table 1.9. Another equally important factor is that high-pH mashes (say, above 5.5) tend to lead to dull malt flavor that lacks definition (Narziss, 1992). Hop flavors are also negatively affected.

Biological acidification is an old method that is experiencing a rebirth. For example, L. Narziss in his article on brewing technology of the future cited biological acidification as a major item (Narziss, 1997). It was originally proposed as a method for controlling the mash pH that is natural to beer. It was targeted at brewers, such as those in

Table I.9	Amylase Activity at 60 °C	
	pH	**Activity (%)**
	4.8	98
	5.0	99
	5.2	100
	5.4	95
	5.8	85
	6.2	65
NOTE: Adapted from Hind (1950).		

Germany, whose compliance with purity laws excluded synthetic chemical additives. Today, many other attributes are associated with this procedure, including a cleaner aftertaste as well as a better defini-tion of the malt and hop flavors (Oliver and Dauman, 1988a, 1988b; Bach and Fersing, 1997; Narziss, 1998). Personal test brews using this procedure support these claims.

Crucial to this procedure is the use of special microbes whose only products are lactic acid. *Lactobacillus delbruckii* is a classical choice (De Clerck, 1957), but *Lactobacillus amylolyticus* is finding favor (Kunze, 1996). It must be emphasized that random contamina-tion of the mash—for example, by overnight mashing—can lead to erratic results. Contamination can introduce undesirable microbes (e.g., butyric acid bacteria). These microbes are eliminated during wort boiling; however, their products have a pernicious way of surviv-ing (at least in part) into the finished beer.

The best results in my test brews came from preparing the sour mash separately from the main mash. The sour mash, if pasteurized, can be stored at room temperature for weeks (and possibly longer). On the brew day, the sour mash is then added incrementally until the desired pH (5.2 to 5.4) is achieved. With specialty beers, e.g., Belgian or German "white" beers, the pH of the mash is reduced to the range

4.0–4.5. It is interesting that the same is being done in Germany with low-alcohol beers (Narziss, 1998). The strong buffering agents in the mash will keep this pH stable once it is established.

When starch conversion is complete, the mash is separated from the grains, a process that leaves residual extract in the grains. To remove that residue, the grains are washed (sparged) with hot water. From the point of view of efficiency, i.e., obtaining the most extract, the more sparge water, the better. Excess volumes can be reduced by appropriate evaporation in the kettle boil. However, it is rare in brewing for efficiency and beer quality to be in harmony. This adage is particularly the case in sparging, because in addition to removing extract, the sparge process also extracts a collection of undesirable compounds, traditionally called *herbstoffe* (astringents). In malt, these come from husk-derived phenols (tannins), silica ash, and other materials that themselves have astringent flavors. In addition, these undesirable elements contribute to beer staling and haze. Raw grains, corn more than rice, also contribute astringent compounds (Fix, 1992).

Thus, brewers must seek a compromise between the smoothness and elegance of a finished beer and the amount of extract they obtain. The traditional rule is to use approximately the same amount of water in the sparge as used in the mash.

In beer styles where smoothness is an important characteristic, brewers tend to augment the traditional rule with the following constraint: When the residual extract in the grains approaches 1.0 °P, then the sparge is terminated. Evaporation in the kettle boil can then be adjusted to get the proper volume for the fermenter. In modern practice, the focus has turned to the pH of the wort collected from the sparge because pH increases with the extraction of undesirable astringents. A general rule is to terminate the sparge when the pH of the collected wort increases much beyond 0.1 pH units higher than the mash pH. In any case, it should not exceed 5.5. Application of this general rule usually amounts to a 1–2 °P extract left in the grains.

If the sparge water has an alkalinity in excess of 50 mg/L as $CaCO_3$, then the pH will quickly approach the critical point. High temperatures also increase the extraction of undesirable compounds, the critical temperature being about 77 °C. Thus, best results are generally obtained with a sparge in the range of 74 to 75 °C and with water whose alkalinity is as low as possible (25 mg/L or below).

Wort Boiling

2

Wort boiling is the one step that is used by all brewers, from beginners working with malt extracts to experts working with sophisticated equipment. It is certainly the easiest part of brewing to manage and, until recent years, has not been given a great deal of attention other than the technological issues associated with equipment. It is ironic that from a purely scientific point of view, it is by a wide margin the most sophisticated part of brewing. The complex redox and Maillard reactions that take place make the microbiological systems in the fermentation look like child's play in comparison. Moreover, there is growing awareness that these reactions do affect finished beer flavors in nontrivial ways, and hence, this area of brewing should not be ignored.

The next section discusses hops and focuses on the fundamental issues that have undergone a considerable evolution since the first edition of this book. For a review of all of the varieties available, see Fix and Fix (1997).

The last two sections discuss the need to establish a sufficiently vigorous boil so that sulfur constituents, most notably dimethyl sulfide (DMS), are appropriately reduced, while at the same time, production of certain Maillard products is minimized. Some Maillard products have less than impressive flavors, particularly when they arise from excess thermal loading during the boil. Successful brewing, like gourmet cooking, is very much a balancing act, and these two aspects of wort boiling are but one example of this point.

HOPS

Hops are botanically classified as *Humulus lupulus*. Only the flowers, or cones, of the female plant are of interest to brewers. Their brewing value comes from the resins and essential oils found in the lupulin glands (Neve, 1991; Garetz, 1994).

There are two types of resins. *Hard resins* are those that are insoluble in hexane. In fresh hops, they consist of at most 5% to 6% of the total resin content (measured by weight) although this percentage generally increases as the hop deteriorates while aging in storage. The brewing value of the hard resins is negative.

The *soft resins* are those that are soluble in hexane. They have been subdivided into α-acids, ß-acids, and uncharacterized soft resins.

α-**Acids.** The bulk of the hop bitterness in beer is due to α-acids (Verzeli and de Keukeleire, 1991).These are phenolic-like compounds (i.e., they contain aromatic rings) that consist of humulone and its analogues. The α-acid structure is

α–acid

In this structure, R stands for a side chain that defines the humulone analogue, as follows:

Name	R (= Acyl side chain)
Humulone	$COCH_2CH(CH_3)_2$
Cohumulone	$COCH(CH_3)_2$
Adhumulone	$COCH(CH_3)CH_2CH_3$

Observe that cohumulone has one less carbon atom than the other analogues, which among other things means that after isomerization (a chemical process in which an organic compound is transformed into another organic compound—an isomer—having the identical chemical composition and molecular weight, but a different structure), cohumulone is more soluble than the other isomerized analogues. Isomerization of cohumulone is described later in this section. It is now generally acknowledged that cohumulone gives a harsher bitter than the other analogues (Rigby, 1972). Hops like Brewers Gold, Clusters, Eroica, and Galena invariably have cohumulone levels in excess of 35% of the total α-acids and are indeed known for their very pungent bitter. In contrast, noble varieties like Hallertau, Saaz, and Tettnang always have cohumulone levels below 25%.

It was once feared that high cohumulone levels were endemic to high-α hops, i.e., hops like Eroica and Galena whose α-acid levels are in excess of 10%. This concern was initially countered by brewers using Columbus, a high-α and moderate–cohumulone hop that recalls Cascade and Centennial, yet because of its fine bitter and aroma, has been widely used by ale brewers both as a kettle and dry hop. Even more striking is Horizon, a new high-α hop whose cohumulone levels are at or slightly below those in noble varieties. Many lager brewers who have a long tradition of commitment to low-α hops are showing interest in this variety.

How a hop is used is also important. Because of the high solubility of cohumulone, brief contact times with wort can lead to undesirably high cohumulone levels even with noble hops. Late hopping in a whirlpool is an example (Mitter, 1995) that is discussed later in this chapter.

A point that cannot be overemphasized is that the characteristics of hops, like grapes, depend on where they are grown. The dramatic difference between the Saaz varieties grown in the United States and those from the Saaz region in Bohemia is but one example of this point. This assertion is not to imply that the former is an inferior hop, but rather that it is different from its European counterpart and should be evaluated on its own merit.

ß-Acids. The ß-acids have the following structure:

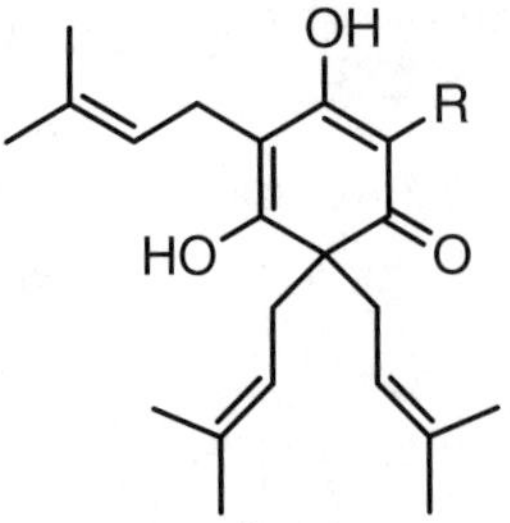

ß-acid

Again, the R stands for a side chain. In ß-acids, the R is one of the following three analogues:

Name	R (= Acyl side chain)
Lupulone	$CH_2CH(CH_3)_2$
Co-lupulone	$CH(CH_3)_2$
Adlupulone	$CH(CH_3)CH_2CH_3$

These ß-acid compounds generally do not play a major roll in fresh hops. However, as hops age they oxidize to more soluble compounds that form an important part of the nonhumulone bitter of beer (Wackerbauer, 1993). The desirability of these compounds depends very much on their concentration and on the type of beer being brewed.

The ratio of α-acids to ß-acids is often cited—along with the percentage of cohumulone—as an important quality indicator. Lager brewers in particular are generally sensitive to this index; however, explanations showing why this index should be important have not been developed. All of the noble varieties have α/ß ratios in the range 0.8 to 1.2. Varieties like Chinook and Eroica, on the other hand, have ratios in excess of 2.5 and generally near 3. It is also striking that Horizon, a high-α hop that is attracting the attention of many lager brewers, has an α/ß ratio below 2. This is the lowest ratio among the major high-α varieties.

Hops with poor storage stability can lose significant amounts of their α-acids in just one year. For example, the α-acid level of refrigerated Cascade pellets dropped from 7.6% to 4.6% after 12 months (Peacock et al., 1981). In the same study, the α-acid level of Hersbrucker pellets dropped from 7.4% to 4.7%. Other varieties showed lower rates of α-acid reduction.

The end products of the oxidation of humulones during storage (and in the kettle boil) comprise a complex array of different compounds (Verzeli and de Keukeleire, 1991). The reaction involves the uptake of oxygen atoms, the following being typical:

α-acid   Oxidized product

The acyl side chain R can also be cleaved via oxidation, giving valeric, butyric, and 2-methyl butyric acids via the mechanisms described in chapter 4. All of these have unmistakable "cheesy" tones.

ESSENTIAL OILS

Although the resins provide the primary hop bitterness, the essential oils are responsible for the overall hop presence, particularly the hop aroma. The essential oils include hydrocarbon components, oxygen-bearing components, and sulfur-containing components.

Hydrocarbon components. The hydrocarbon components form slightly less than 75% of the essential oils (Kunze, 1996) and consist of a monoterpene (e.g., $C_{10}H_{16}$) or a sesquiterpene (e.g., $C_{15}H_{24}$). Terpenes are a structurally diverse family of compounds with carbon skeletons like those of humulene and myrcene, the two main hydrocarbon compounds that have been identified in beer. Their structures are shown here.

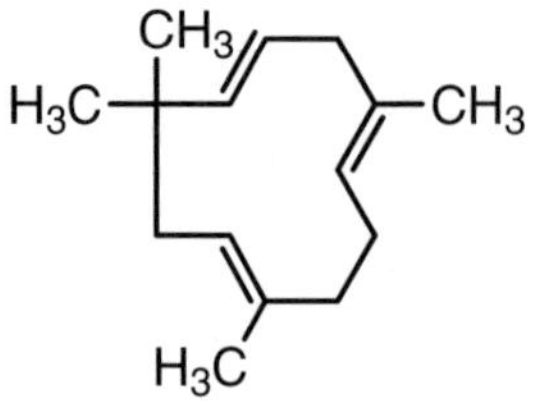

Humulene

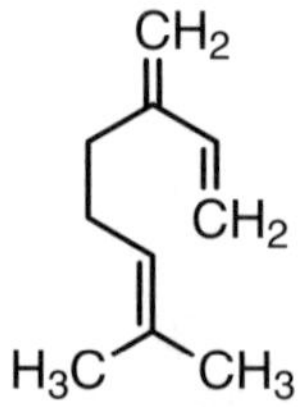

Myrcene

Humulene has a delicate and refined flavor that is often described as "elegant." Myrcene has a greater flavor intensity, frequently characterized as "pungent." Because of its structure, myrcene is usually classified as a monoterpene, whereas humulene is classified as a sesquiterpene. Two other sesquiterpenes of interest are farnesene and caryophyllene:

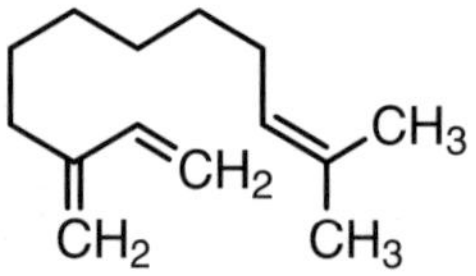

Farnesene

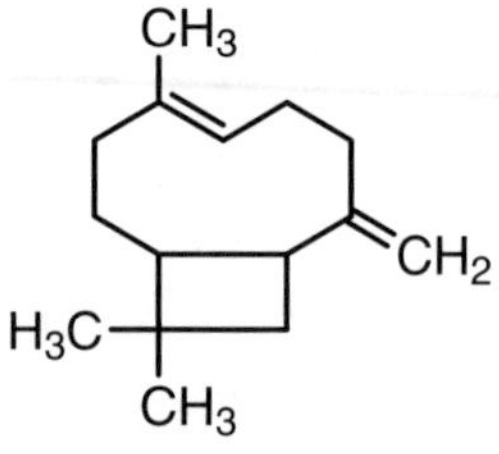

Caryophyllene

Distinctions between noble and more aggressively flavored hops can be seen in their content of essential oils. The noble hops typically have very high sesquiterpene/monoterpene ratios, generally 2.5–4. On the other hand, the pungent hops are typically dominated by myrcene, which can constitute almost 60% of their essential oils (thus yielding a sesquiterpene/monoterpene ratio near 0.67).

Another widely cited index is the H/C ratio, i.e., the ratio of humulene to caryophyllene. The H/C ratio of noble varieties will generally exceed 3.5, although there are variations. Thus, this index is probably a better discriminator among noble hops than a discriminator in general. The same is true for the farnesene level. For example, Saaz-type hops (Saaz, Tettnang, Spalt) tend to have much higher values than the Hallertau types (Peacock, 1992). A summary of the various indices cited above for selected hop types is given in Tables 2.1 to 2.3. These data were taken from Garetz (1994), Peacock (1992), and Haunold (1998).

The data cited in Tables 2.1–2.3 are for fresh-leaf hops. Oxidation will alter their composition, especially the essential oils, as described later in this chapter. Also, heat treatment such as pelletization can alter hop composition. For example, high-farnesene hops like Saaz and Tettnanger tend to lose the farnesene oil component during pelletization. Such farnesene reduction is likely the main reason why there are clear differences in the character of beer aroma depending whether leaf or pelletized versions of these hops are used.

Oxygen-bearing components. The oxygen-bearing components form approximately 25% of the hop essential oils (Kunze, 1996). Nevertheless, in selected varieties, these oil constituents can play a major role in defining the hop's basic flavor characteristics. Two alcohols in this category, namely linalool and geraniol, have been extensively studied (Peacock et al., 1981). In isolation, linalool has a pleasing hoppy smell, not as elegant as humulene, but quite attractive. Geraniol, in isolation, tends to recall floral or herbal cheap perfume. Nevertheless, there is a synergy between these oils. Very floral or herbal hops like Amarillo, Cascade, Centennial, and Columbus tend to have very high concentrations of both oils, which is likely responsible for their very special aroma (Peacock et al., 1981).

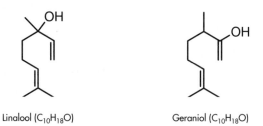

Linalool ($C_{10}H_{18}O$)   Geraniol ($C_{10}H_{18}O$)

Selected German and Czech Hops

Table 2.1

Name	α	α/β	CoH	MYR	HUM	F/H	H/C	Comments
Czech Saaz	3–4.5	1	20–25	30–35	25–30	1.5–2	3.5–4	Traditional noble hops
Hallertauer Mittelfruh	3–4.5	1	20–25	30–35	35–40	0	3.5–4	Traditional noble hops
Hersbrucker	3–5.5	1	20–25	35–40	30–35	0	3–3.5	Traditional noble hops
Spalt	4–5.5	1	25–30	15–25	15–20	1.5–2.0	1.5–2	Traditional noble hops
Tettnanger	3.5–5.5	1	20–25	30–35	20–25	1.5–2	3.5–4	Traditional noble hops
Hallertauer Tradition	5–7	1–1.5	25–30	20–25	45–55	0	.3–3.5	New
Spalter Select	5–6	1–1.5	20–25	15–25	15–25	1.5–2.0	2–2.5	Low-α hops
Northern Brewer	7–10	2–2.5	30–35	30–35	25–30	0	3–3.5	Other
Perle	7–10	1.5–2.0	30–35	45–55	30–35	0	2.5–3	Other

NOTE: α = α-acid level (% of the total acids), α/β = ratio of α-acid to β-acid, CoH = cohumulone (% of the total α-acids), MYR = myrcene (% of the total essential oils), HUM = humulene (% of the total essential oils), F/H = farnesene/humulene ratio, and H/C = humulene/caryophyllene ratio. Data from Garetz (1994), Peacock (1992), and Haunold (1998).

Selected U.K. Hops

Table 2.2

Name	α	α/β	CoH	MYR	HUM	F/H	H/C	Comments
East Kent Golding (EKG)	4–5.5	1.5–2.5	20–25	45–50	25–30	0	3.5	Traditional U.K. aroma hops
Fuggles	4.5–5	1.5–2	20–25	25–30	35–40	1.5–2	3.3	Traditional U.K. aroma hops
First Gold	7–10	1.5–2	30–35	25–30	20–25	0.1–.2	3–3.5	New EKG replacement
Admiral	11.5–14.5	2–3	40–45	40–50	25–30	0.1–.15	3–3.5	High α-hops
Phoenix	8–11.5	1.5–2	30–35	25–35	25–35	0	2.5–3	High α-hops
Wye Target	10–13	2.2	35–40	60–65	10–15	0	2.4	High α-hops

NOTE: α = α-acid level (% of the total acids), α/β = ratio of α-acid to β-acid, CoH = cohumulone (% of the total α-acids), MYR = myrcene (% of the total essential oils), HUM = humulene (% of the total essential oils), F/H = farnesene/humulene ratio, and H/C = humulene/caryophyllene ratio. Data from Garetz (1994), Peacock (1992), and Haunold (1998).

Table 2.3

Selected U.S. Hops

Name	α	α/β	CoH	MYR	HUM	F/H	H/C	Comment
Crystal	3.5–5	0.5–1	20–25	40–50	20–30	0	3.5–4	Low-α hops
Liberty	3–5	1–1.5	25–30	35–40	35–40	0	3.5–4	Low-α hops
Mt. Hood	5–8	1	20–25	55–65	15–25	0	2–2.5	Low-α hops
Santiam	6–7	0.8–1.0	15–20	30–35	20–25	0.5	3.2–3.5	Low-α hops
Tettnanger	4–5	1–1.5	20–25	35–45	20–25	1.0–1.5	3–3.2	Low-α hops
Cascade	4.5–7	1	35–40	45–60	10–15	1.0–1.5	3.5–4	Moderate-α hops
Amarillo	4.5–8	1	30–40	40–50	10–20	1–1.5	3.5–4	Moderate-α hops
Centennial	9.5–11.5	2.5–3	30–35	45–55	15–20	0	2.2	Moderate-α hops
Northern Brewer	8–10	2–2.5	20–30	50–60	20–30	0	3–3.5	Moderate-α hops
Brewers Gold	5.5–8.5	2–2.5	40–45	60–65	10–15	0	2–2.5	Pungent; moderate-α hops
Cluster	5.5–8.5	1–1.5	35–45	45–55	15–20	0	2–2.5	Pungent; moderate-α hops
Columbus	14–16	3–4	25–30	25–30	15–25	0	2–3	Special high-α hops
Horizon	11–13	1.7	15–20	40–50	10–15	0.3	1.5–2	
Chinook	12–14	3.5–4	30–35	35–40	20–25	0	2–2.5	High-α hops
Eroica	11–13	2.5–3	35–45	55–65	0	0	0	High-α hops
Galena	12–14	1.5–2.0	40–45	55–60	10–15	0	3	High-α hops
Nugget	12–14	2.5–3	25–30	50–60	10–20	0	2	High-v hops

NOTE: α = α-acid level (% of the total acids), α/β = ratio of α-acid to β-acid, CoH = cohumulone (% of the total α-acids), MYR = myrcene (% of the total essential oils), HUM = humulene (% of the total essential oils), F/H = farnesene/humulene ratio, and H/C = humulene/caryophyllene ratio. Data from Garetz (1994), Peacock (1992), and Haunold (1998).

As noted, the hydrocarbons undergo alterations of their oil structure during storage. For example, one study reported that the myrcene level of Cascade hop pellets decreased from 329 mg/L to 7 mg/L after 12 months of refrigerated storage (Haley and Peppard, 1983). For this reason, many now believe that the oxidized products of hop oils are actually more important than primary hydrocarbons like humulene and myrcene.

Sesquiterpenes such as those listed below are known to be important oxidation products. Most studies show that they consistently increase during hop storage as the primary hydrocarbons decrease (Haley and Peppard, 1983). Five sesquiterpenes derived from humulene are humuladienone, humelene, epoxide I, humelene epoxide 2, humelol, and humulenol. Their structures are key shaped:

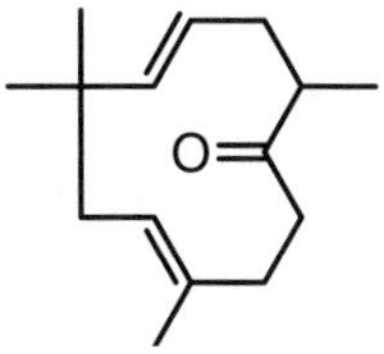

Humuladienone

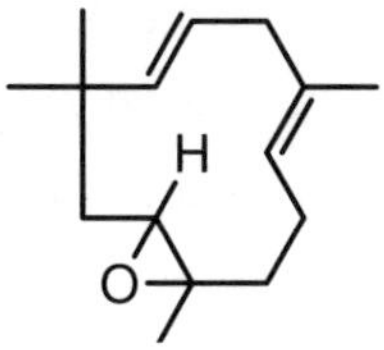

Humulene epoxide 1

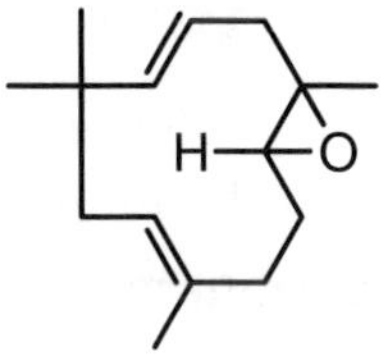

Humulene epoxide 2

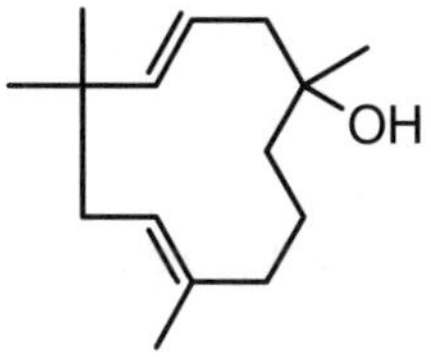

Humelol

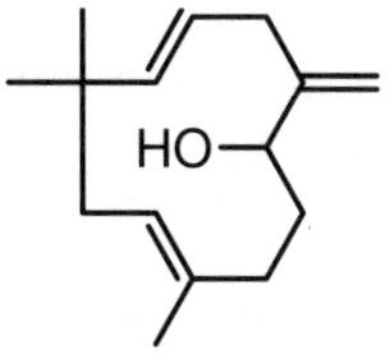

Humulenol

Oxidation of myrcene leads to the analogous products myrcene epoxide and myrcenol:

Myrcene epoxide   Myrcenol

The oxidation products from humulene tend to have sensory characteristics that recall hay or sagebrush; sometimes the term *grassy* is used. Flavor panels tend to be sharply divided over the desirability of such flavors (Stucky and McDaniel, 1997). The products derived from myrcene, on the other hand, retain the pungency of their precursor, but at a slightly lower intensity.

Sulfur-containing components. Beers containing excessive sulfury tones are generally regarded as defective. Thus, a good deal of attention has been given to the origin of these flavors. It is believed that malt and the fermentation are the major sources of sulfury flavors, but hops are known to contribute to them as well.

Of the various mechanisms that are possible, two are well understood. One is the "light-struck phenomenon" described by S. Bickham (1998). Light of wavelength from 400 to 500 nm can cause photochemical rearrangement of hop resins, resulting in a sulfury mercaptan compound with a pronounced skunky character. The other mechanism arises from excessive use of sulfur compounds in the field to protect the hops from fungus infestation before harvest. Most of the residual sulfur, however, is removed in the kiln (i.e., during hop drying after harvest), and residual sulfur in hops added early in the kettle boil is generally evaporated and not passed on to the finished beer. High-sulfur hops used in late kettle additions or dry hopping can still produce unacceptable levels of thioesters (*thio* means sulfur-bearing; esters are a class of typically aromatic organic compounds). High thioester production became a serious problem for brewers in England in the 1970s. However, greater caution has been used in fungicidal spraying since that time (Seaton et al., 1981).

Aside from these well-understood mechanisms, there is also the poorly understood role of sulfur-based radicals (i.e., parts of complex compounds) that are natural to hop constituents. Hop oils in particular are known to contain various thiols, sulfides, thioesters, thiophenes, and episulfides. Their concentrations are small, yet so are their flavor thresholds. As a consequence, their overall relevance to beer is yet to be determined.

HOP TRANSFORMATIONS DURING WORT BOILING

Isomerization of hop resins. The α-acids in hops are poorly soluble in wort; under boiling conditions, however, the α-acids are rearranged to more soluble iso-α-acids (iso- indicates that the compound is an isomer of another compound). This transformation is called isomerization. As an example, consider the humulone structure (for purposes of discussion, the six carbon atoms in the phenolic ring are numbered):

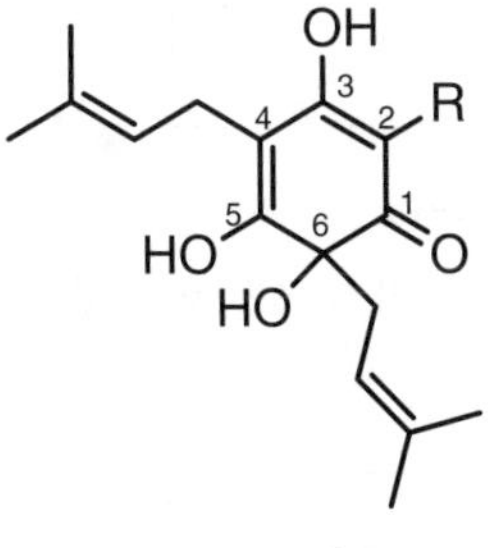

Humulone

Under boiling conditions, the bond between the first and sixth carbon atoms is broken, and a new bond is formed between carbon atoms one and five. The result is isohumulone:

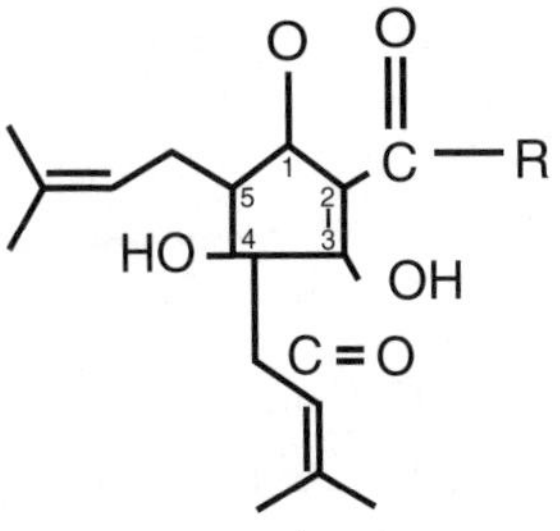

Isohumulone

Similar transformations take place for the humulone analogues cohumulone and adhumulone.

Not all of the α-acids added to the brew kettle will be converted to iso-α-acids. In addition, some of these hop resins are lost during both the hot break and the cold break. It is customary to define the efficiency of this process in the following way as a kettle utilization rate (KUR):

$$\% \text{ KUR} = \frac{\text{amount of iso-}\alpha\text{-acids in wort}}{\text{amount of }\alpha\text{-acids added to kettle}} \times 100\% . \quad (2.1)$$

Kettle utilization rates vary with many factors. However, for a given brewing configuration, it is the hop type (whole, pellet, or extract), contact time, temperature vs. pressure profile, wort gravity, and pH that cause batch-to-batch variations (Fix and Fix, 1997). With these factors under control, brewers can make accurate estimates of their KUR from measurements of test brews. After that, the wort iso-α-acid content can be directly computed from the α-acid content of the hops used. In metric units (i.e., SI—International System of Units), this calculation is trivial. For example, using 200 g of 5% α hops per hectoliter is equivalent to an α-acid level of

$$0.05/\text{hL} \times 200 \text{ g} = 10 \text{ g/hL}$$

and, through standard conversion of units,

$$10 \text{ g/hL} \times 10^{-2} \text{ hL/L} \times 10^3 \text{ mg/g} = 100 \text{ mg/L}.$$

If the KUR is 25%, then the iso-α-acid level of wort is

$$100 \text{ mg/L} \times 0.25 = 25 \text{ mg/L}.$$

For those working in oz./bbl or oz./gal., it is easiest to convert values in these units to metric units via the following conversion relationships:

$$1 \text{ oz./gal.} = 748.5 \text{ g/hL}$$
$$1 \text{ oz./bbl} = 24.15 \text{ g/hL.}$$

Then do the rest of the calculation in metric units.

The term *bitterness unit* (BU) typically refers to a number measured by a spectrophotometer (see *Methods of Analysis,* American Society of Brewing Chemists, 1987). It measures selected soft resins as well as iso-α-acids, and the result is typically 5%–10% higher than the actual iso-α-acid level. The iso-α-acid level in finished beer is typically less than that of wort. For example, losses in the fermentation occur when hop constituents get trapped in the bottom yeast layer (Garetz, 1994). These losses, however, tend to be constant in any given brewing configuration, so they can be estimated from measurements of test brews. Garetz (1994) also has given some useful rules for determining approximate losses.

It must be emphasized that the perception of hop flavor (taste and smell) is far too subtle to be captured by one single number. It is not uncommon for one beer to be rated as having a milder hop bitter than another in blind tastings, even though the former has a higher BU (see, for example, Preis and Mitter, 1995). Moreover, there is absolutely no correlation between BUs and the quality of the hop flavor. Thus, BUs should be seen as a rough estimate of the overall intensity of a beer's hop profile, but at the end of the day, the human palate reigns supreme as the ultimate arbitrator.

Oxidation and/or polymerization of hop resins and oils. As noted above, the reaction mechanisms in wort boiling are by far the most sophisticated to be found in brewing. This fact is illustrated by complicated transformations involving humulones and deoxyhumulones (Shannon et al., 1978; these transformations have been further elaborated on by Verzeli and Keukeleire, 1991). Ironically, the overall relevance of these mechanisms is still unresolved.

There are, however, some established facts about transformations of cohumulone. As noted earlier, isocohumulone is the most soluble of the humulone analogues. Moreover, it is ultimately transformed into milder products. For example, M. Mitter (1995) compared the isocohumulone concentrations of two brews. The brews were identical in composition and brewing method except for one thing: the hops were added with the first wort (first-wort hopping), but in the second brew, the hops were added at the very end of the boil. A low–cohumulone hop (Tettnanger) was used in both brews. It was found that the isocohumulone level of the first-wort-hopping brew was 20% lower, and blind tasting of the finished beers showed a clear preference for it. Additional studies have shown that actually as little as 15 minutes of wort boiling will reduce the isocohumulone level to within 5% of the value of first-wort hopping.

Transformations of the hop oils are even more important as far as finish beer flavor is concerned. The lovely aroma that occurs after hops are added to boiling wort suggests that perhaps there is a considerable loss of hop oils. Moreover, it would seem that the most desirable components are the first to go. This problem has led to a distinction between "bitter hops" and "aroma hops." The former are used solely for their α-acids and are added early into the boil. Aroma hops, on the other hand, are valued for their fine gastronomic qualities and are usually used in late hopping. De Clerck (1957) questioned these assumptions. He asserted that beer always retains the smell of hops no matter when they were added.

Direct evidence of this assertion is given by Preis and Mitter (1995). They compared the two extremes—namely, a late-hopped beer (the "reference" beer) and a first-wort-hopped beer (FWH)– by using both chromatography and blind tasting with a professional taste panel. The comparison was done at two breweries (A and B), and the procedures described in Table 2.4 were followed. The chromatographs of the reference and FWH brews were dramatically different (see Figs. 1 and 2 in Preis and Mitter, 1995). They suggested that there was not so much of a loss of hop oil in the FWH brew, but rather a different spectrum of oil components. There was, however, a

Table 2.4	**Hop Additions**	
	Brewery A	**Brewery B**
Hop type	Tettnang and Saaz pellets	Tettnang pellets
Total α-acid added	13 g/L	13 g/L
Reference brew		
Middle addition	63%	48%
Late addition	37%	52%
FWH brew		
First wort	37%	52%
Middle addition	63%	48%

Table 2.5	**Iso-α-Acid Recovery**	
	Brewery A	**Brewery B**
Wort iso-α-acid		
Reference	47.6	32.6
FWH	55.0	44.8
Beer iso-α-acid		
Reference	40.9	27.4
FWH	42.7	35.1

noticeable drop in the oxygen-bearing components of the hop aroma substances (see Table 2.5). In addition, there was a predictable increase in the iso-α-acids recovery with FWH (see Table 2.4).

What is striking is the strong preference that the professional panels of both breweries showed for the FWH beer. The sensory assessments found that the FWH beer had "a finer and rounded hop aroma" with a "finer and more pleasant" bitter in spite of the higher iso-αacid concentration. The latter conclusion could possibly be explained by the lower cohumulone content. The former came as a surprise to the investigators, and they conjectured that the reduction

Table 2.6	**Hop-Oil Concentrations (μg/L)**	
	Brewery A	**Brewery B**
Linalool		
Reference	29.0	34.1
FWH	8.1	6.4
Geraniol		
Reference	18.8	14.6
FWH	10.7	13.7
Humulene epoxide		
Reference	32.7	10.8
FWH	19.6	9.8

Table 2.7	**Results of the Professional Tasting Panel**	
	Brewery A	**Brewery B**
No. of tasters	12	13
No. passing triangle test	11	12
Preference for FWH beer	8 to 3	11 to 1

in the oxygen-bearing components of the hop oils (Table 2.6) may be the reason. The results of the professional tasting panel in the Preis and Mitter (1995) study are shown in Table 2.7.

After the *Brauwelt* study (Preis and Mitter, 1995) was published, FWH was tried by a large number of homebrewers. Some came to the same conclusions as those in the Brauwelt study, but not everyone agreed (Home Brewers Digest [hbd.org]—Electronic Discussion Group on Beer), which is not very surprising given the highly hetero-geneous nature of homebrewing. Experimentation with FWH has also been done by commercial craft brewers. Preliminary results have been favorable; however, additional research is needed into how well the procedures work with the type of ales that are brewed by craft brewers. The *Brauwelt* study was confined to Pilseners, as were follow-up studies (Fix, 1997).

As noted in chapter 1, dimethyl sulfide (DMS) is the most important sulfur constituent in beer. Aside from bacterial action, the most common reaction system is production of DMS via heat in malt kilning, wort boiling, and wort cooling. The DMS precursor of greatest interest is S-methyl methionine (SMM), which, during heating, breaks down to DMS in the following way:

$$H_3C - S - (CH_2)_2CH(NH_2)C(=O)OH \quad \overset{H_3C}{\underset{H_3C}{|}} \rightarrow \quad H_3C - S - H_3C \; + \; (CH_2)_2CH(NH_2)C(=O)OH$$

SMM   DMS

Breakdown of SMM to form DMS

Even SMM itself is not neutral to finished-beer flavor. For example, free SMM can bond with heterocyclics to form some rather nasty compounds (see the next section).

SMM is formed in malt during germination. Moreover, as the nitrogen content of SMM suggests, its formation closely follows protein levels and protein modification. As a consequence, high-protein malts (e.g., those made from six-row varieties) tend to have higher SMM levels than low-protein malts. For example, English pale malts tend to have 1 to 2 µg of SMM per gram of malt, whereas some North American six-row varieties can run as high as 8 to 10 µg of SMM per gram of malt.

As the malt is dried during the kilning, some SMM is lost by heat decomposition into DMS and other products. These are removed in later processing and are unimportant to DMS levels in the finished wort. It is for this reason that malts produced at high kilning temperatures (e.g., ale and color malts) have lower SMM levels than lager malts.

During wort boiling, SMM is converted to DMS via a first-order reaction mechanism. Under isothermal conditions, this reaction is well approximated by a linear ordinary differential equation of the form

$$-dx/dt = cx,$$ (2.2)

where x is the SMM concentration at time t and c is a positive constant. Equation 2.2 states that the rate of reduction of SMM (i.e., the negative of the derivative dx/dt) is proportional to the instantaneous concentration x. It is customary to write this equation in terms of the half-life H of SMM. By using the concept of half-life, the constant C can be written

$$C = \ln(2^{1/H}),$$ (2.3)

where $2^{1/H}$ represents 2 taken to the power of $(1/H)$ and ln indicates the natural logarithm (Hughes-Hallet et al., 1998). The solution for the concentration x at any time t is

$$x(t) = x_0 \times 2^{-t/H},$$ (2.4)

where x_0 is the initial SMM concentration. Careful measurements have shown that the half-life H of SMM is 40 min for an isothermal boil at $T = 100\ °C$ and ambient pressure.

To illustrate the implications of this model, suppose one is using malt that averages 5 µg of SMM per gram of malt, and suppose that the malt concentration is 20 kg/hL = 200 g/L (which is typical of 12 °P wort). If any DMS produced during mashing is ignored, the SMM content of wort at the start of the boil is

$$5\ µg/g \times 200\ g/L = 1000\ µg/L.$$

Suppose a boil at 100 °C for $t = 90$ min is used. Then

$$1000\ µg/L \times 2^{-(90\ min/40\ min)} = 1000 \times 2^{-2.25} = 1000 \times 0.21 = 210\ µg/L$$

of SMM survives the boil, and the rest is transformed to DMS, which means

$$1000 \ \mu g/L - 210 \ \mu g/L = 790 \ \mu g/L$$

of DMS was created. Reasonable kettle ventilation allows the DMS so formed to be removed with the vapor. The residual, i.e., 210 µg/L of free SMM, is considered as DMS potential, since the transformation SMM → DMS can still occur after wort boiling.

However, the most obvious place where this transformation occurs is during wort cooling, particularly if this is done in a closed environment. The half-life of SMM (H_{SMM}) is approximately related to temperature T by an inverse relationship. As the temperature falls below 100 °C, the H_{SMM} will increase from its value of 40 min at 100 °C. If $T = T(t)$ denotes the temperature profile over time t, then measurements indicate that the half-life at any given time $H(t)$ obeys (to a first approximation) the inverse relationship

$$H(t) = 4000/T(t) \qquad (2.5)$$

for 50 °C < $T(t)$ < 100 °C. At temperatures below 50 °C, the number 4000 must be replaced with a function of time t, which rapidly increases below 40 °C. This implies that the changes in the SMM concentration over time satisfy a time-dependent differential equation (equation 2.2), where C is now a function of time via equations 2.3 and 2.5. An approximate solution is given by

$$x(t) = x_0 \times 2^{-k(t)} , \qquad (2.6)$$

where

$$k(t) = t \times T(\tfrac{1}{2})/4000 . \qquad (2.7)$$

Observe that if $T = 100$, then equations 2.6 and 2.7 reduce to equation 2.4 and $H = 40$ min.

A couple of numerical examples may clarify this topic. Suppose the wort cools from 100 °C to 10 °C in $t = 60$ min. Taking

$$T(\tfrac{1}{2}) = (100 \ °C + 10 \ °C)/2 = 55 \ °C ,$$

then

$$k(t) = (60 \ \text{min} \times 55 \ °C)/(4000 \ \text{min·}°C) = 0.825.$$

This result means that of the 210 µg of SMM per liter remaining after the 90 min boil, the concentration of SMM left after cooling to 10 °C in $t = 60$ min is

$$210 \text{ µg/L} \times 2^{-0.825} = 210 \text{ µg/L} \times 0.564 = 118 \text{ µg/L},$$

and the concentration of DMS created during the cooling is

$$210 \text{ µg/L} - 118 \text{ µg/L} = 92 \text{ µg/L}.$$

Practically all of the DMS will be dissolved in wort if a closed cooling system is used. Note that this concentration is slightly over three times the DMS threshold of 30 µg/L.

On the other hand, if the cooling time is cut in half to $t = 30$ min, then

$$k(t) = (30 \text{ min} \times 55 \text{ °C})/(4000 \text{ min·°C}) = 0.4125,$$

and so

$$210 \text{ µg/L} \times 2^{-0.4125} = 210 \text{ µg/L} \times 0.752) = 157 \text{ µg/L}$$

of SMM remains, giving a dissolved DMS concentration of

$$210 \text{ µg/L} - 157 \text{ µg/L} = 53 \text{ µg/L} .$$

Chapter 3 shows that fermentation at ambient temperatures will cause an approximately 50% reduction in both SMM and DMS due to CO_2 evolution. For cold fermentations at 10 °C, however, SMM and DMS will only be reduced by 25%. Thus for lager brewers, the first scenario will have an objectionable concentration of DMS according to Kunze's rule of no more than 60 µg/L (see chapter 1), but the short ($t = 30$ min) cooling procedures will likely lead to a normal DMS level between 30 to 50 µg/L. Nevertheless, many lager brewers worry about the SMM residuals in both cases, and as a consequence, take extra steps. The largest brewery in the United States sprays the hot wort with nitrogen gas, which removes virtually all DMS and SMM. However, the procedure generally removes some desirable compounds as well, leaving a finished beer with an inordinately low malt profile. Another procedure that is used for reducing (as opposed to eliminating) SMM and DMS levels consists of holding wort at 90–95 °C, allowing for maximum ventilation. Equations 2.6 and 2.7 allow one to determine the time period required.

For ales, these sulfur-compound issues tend to disappear. For example, ale malt with only 1 μg of SMM per gram of malt is common. This low initial concentration means that the SMM level at the start of the boil is

$$1 \ \mu g/g \times 200 \ g/L = 200 \ \mu g/L.$$

This initial SMM concentration leaves a SMM residual of

$$200 \ \mu g/L \times 0.21 \ = 41 \ \mu g/L$$

at the end of the boil. As noted above, 50% of the sulfur-containing compounds (be they SMM or DMS) is removed in the fermentation. Thus, the finished beer will have subthreshhold levels of DMS no matter how the wort is cooled. Indeed, any hint of DMS in ales is likely from technical brewing errors, most notably contamination (see chapter 3). This fact probably explains why many ale aficionados react very negatively to DMS in any type of beer.

NONENZYMATIC BROWNING

The Maillard reactions discussed in chapter 1 in the context of malting also occur in wort boiling. Schematically, the chemical process involved is represented as follows:

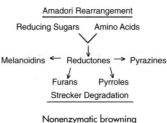

This process, which is sometimes called nonenzymatic browning (NEB), is remarkably efficient. For example, the carbohydrate profile is modified only in trivial ways, and amino acids used in the process account for only 1% of the total, yet the wort color can typically double. The heterocyclic compounds formed in the process can be assigned to one of the three following levels of complexity:

- Level 1—Simple melanoidins
- Level 2—Sulfur-bearing heterocyclic compounds
- Level 3—Strongly flavored heterocyclics

As the term *reductones* suggests, they are natural antioxidants. These compounds are found in abundance in well-executed decoction mashes. Of all wort reductones, the simple melanoidins in their reduced state are the most desirable. They have an attractive "toasty" flavor in amber and dark beers. Similar results can be achieved with worts produced from infusion mashes if boiling is properly conducted. On the other hand, excessive thermal loading (e.g., uncontrolled high-temperature and/or high-pressure boils) can transform the simple melanoidins into less desirable heterocyclics such as described next (Narziss, 1992).

The second level of complexity consists of sulfur-bearing heterocyclics. Heterocyclic and methionine products have a flavor that strikingly recalls cooked cabbage. The potential presence of these compounds in beer is responsible for the sensitivity of many brewers to free SMM in wort. Nevertheless, it is excessive heat treatment— not inadequate boiling—that is responsible for their formation (Heath, 1988).

The third and final level of heterocyclic compound complexity is generally regarded as giving off-flavors except for very special beer styles. Pyrazines are the most studied class of nitrogen heterocyclics formed via NEB (Heath, 1988), and it is also the most relevant to wort that has been subject to heavy thermal loading. The alkyl pyrazines are common; their structures were described in chapter 1. An example commonly found in wort produced by boiling under pressure is 2-acetonyl pyrazine, which has a sharp toasted or burnt flavor. Its structure is shown here:

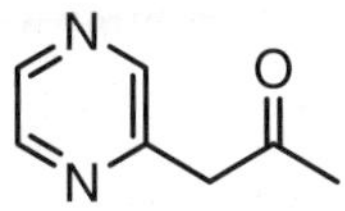

Structure of 2-acetonyl pyrazine

Pyrroles constitute another class of nitrogen-bearing heterocyclics relevant to wort boiling. An important example is 2-formyl pyrrole, which has a sweet-corn-like flavor. It has the following structure:

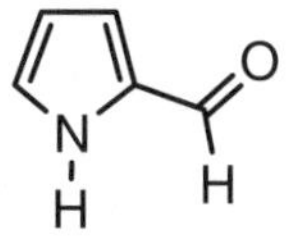

Structure of 2-formyl pyrrole

It is often confused with DMS, yet the mechanisms responsible for their formation are completely different. The heterocyclic increases with increasing thermal loading, whereas increasing thermal loading encourages the reduction of DMS by vaporization (Herrmann et al., 1985).

Furanones are the most characteristic of the level-three heterocyclics and have flavors that have been variously described as burnt, caramel-like, nutty, or smoky. Furan derivatives are also often found in bread and coffee (Heath, 1988).

Because of the potential for heterocyclic-compound formation, wort boiling poses a dilemma for brewers. Thermal loading on wort should be sufficiently high to expel DMS, yet not so high as to produce undesirable off-flavors. Even for dark beers, it is usually better to use the malt to define the malt flavor profile than to rely on heterocyclics formed in wort boiling. This dilemma is aggravated by the fact that during wort boiling, the system is very far from equilibrium, and hence standard pressure-temperature-volume relationships from equilibrium thermodynamics are not valid. As a consequence, select parameters are used as indicators. These are not fundamental in themselves, but they do tend to track thermal loading within a given system. The most widely used indicator is the percent evaporation that takes place in the boil (Narziss, 1992). With standard boiling systems, a general rule is that the volume reduction be at least 7%. However, it has been shown that evaporation rates above 12% may produce level 2 heterocyclics, leaving vegetal malt tones that are accompanied with some astringency. A wide range of level 2 and 3 heterocyclics is possible once evaporation rates exceed 15%. As already stated, the flavor of the finished beer will determine the extent to which this effect is relevant.

Protein coagulation also occurs during wort boiling and is highly desirable as far as finished beer is concerned. Protein coagulation primarily constitutes the so-called "hot break" and is evidenced just after the boil where clear wort has protein complexes suspended in it. The "cold break" refers to sediment formed after wort cooling. The stability of beers with respect to haze requires the sufficient removal of both cold and hot breaks. Not much in the way of thermal loading is needed to get good breaks (hot and cold) with modern malts. Some studies have concluded that a little as a 2% volume reduction is adequate (Buckee et al., 1982).

Fermentation

3

YEASTS

It was from Louis Pasteur's fundamental studies (Pasteur, 1876) that brewers learned that various unicellular microorganisms were responsible for the fermentation of beer, as well as the spoilage of beer in adverse circumstances. These fungi are quite small, typically micrometers in size (1 micrometer = 1 μm = 10^{-6} m = 10^{-4} cm). Thus, their individual cells are invisible to the naked eye and require a microscope to be detected. Since Pasteur's time, it has become clear that the most important part of a brewer's art is the proper control of adverse microorganisms like bacteria and wild yeasts and the proper care of the good ones like brewing yeasts.

CLASSIFICATION AND SELECTION OF YEASTS

Biologists classify a microorganism such as ale yeasts by first determining the appropriate broad category, called a genus, and then the specific subcategory, called a species, within the given genus. Thus, ale yeasts are *Saccharomyces cerevisiae*, which indicates that this yeast organism belongs to the genus *Saccharomyces* and the species *cerevisiae*. Lager yeasts belong to the same genus but to a different species, as seen by the biological name *Saccharomyces carlsbergenis*, sometimes called *Saccharomyces cerevisiae* vars *uvarum* in other disciplines. In a different genus are yeasts of another kind—*Brettanomyces lambicus*—prized by Belgians for brewing lambic beers.

This classification method has a number of limitations as far as brewing is concerned. First, adverse microorganisms are constantly changing, largely in response to brewers' attempts to control them. Second, even cultured brewing yeasts can perform in dramatically different ways in actual beer fermentations. To avoid these difficulties, this book follows the functional approach used in *An Analysis of Brewing Techniques* (Fix and Fix, 1997). In this approach, a microbe is defined by how it behaves. The key behavioral characteristics are indicated by yeast-cell morphology, flocculation, attenuation, preferred temperature range, and the production of fermentation by-products.

Morphology. A microscope can be used to ascertain the physical appearance of yeast cells. Most have diameters in the range of 5 to 10 μm. Some immature cells can be as small as 3 μm. The cells should be round or oval. Unusual shapes can indicate disorders. For example, yeasts stored without proper nutrients will quickly starve, and in that condition the cells take on highly irregular, elongated shapes. Wild yeasts, e.g., *Saccharomyces pastorianus*, naturally have elongated cells. Cells of wine yeasts, i.e., *S. cervesiae* vars *ellipsoidens*, have a characteristic elliptic shape. It must be emphasized, however, that a simple microscopic examination has limitations. Nonculture strains of various types can be present at levels high enough to affect beer flavors, yet still not be easily detectable during examination under a microscope (Pipes, 1978). Elementary but accurate determination of nonbrewing yeasts requires the techniques described in *Methods of Analysis* (American Society of Brewing Chemists, 1987) or in the work by Fix and Fix (1997); these techniques use specialized media for incubation. This procedure can take several days or even weeks before small numbers of contaminating cells can multiply to a population size big enough for them to be detected (Fix and Fix, 1997).

Flocculation. Yeast flocculation is a genetic characteristic that can change with mutation (Russel, 1995). There are three types of flocculating behavior: strongly sedimenting, powdery, and nonflocculating. In terms of wort attenuation and the reduction of undesirable by-products like diacetyl, yeast cells that remain largely in suspension for the duration of the fermentation invariably give the best results. Many brewers in fact use nonflocculating strains that are removed from the fermented wort by mechanical means (e.g., centrifuging).

Other brewers have a general preference for powdery strains. Strongly flocculating strains are still being used by some brewers; however, the practice of "rousing" (i.e., breaking up the bottom yeast sediment) is usually used in conjunction with these strains.

A change in a yeast strain's flocculation pattern is a cause for concern. First, it could signal mutation. A more common situation is one in which a brewer has a mixed, rather than pure, strain, and in the course of collection and repitching, one strain starts to dominate. For example, "bottom cropping," i.e., collecting yeasts from the bottom sediment, can allow strongly flocculating strains to become dominant.

Attenuation. The yeast cells' ability to attenuate the initial specific gravity of the wort by fermenting the sugars is another genetic property that is subject to mutation. There are also environmental factors that may cause attenuation. For example, some yeasts tend to be highly sensitive to alcohol, and their performance is therefore dependent on the strength of the wort fermented. There are genetically engineered strains that are capable of fully fermenting 18 °P wort (Russel, 1995). Most strains start to change their normal attenuation pattern in the range of 14–16 °P. More commonly, the brewer fails to meet the strain's oxygen demand, which varies considerably from strain to strain. Failure to dissolve an adequate amount of oxygen into the chilled wort can lead to a number of disorders.

To avoid these confounding issues, it is best to measure a yeast strain's intrinsic attenuation in terms of a standardized wort that has been saturated with O_2 before pitching. A common choice is a 12 °P (1.048) wort, which has a real degree of fermentability of 65% or, equivalently, an apparent fermentability of approximately 80%. Both of these measurements can be checked by the classic forced fermentation procedure (Fix and Fix, 1997). Table 3.1 lists the intrinsic abilities of various yeasts to ferment all the sugars in wort.

The manner in which individual yeast strains ferment trisaccharides can give valuable information about the strain in question. For example, normal lager strains will completely ferment raffinose, a minor wort sugar. Ale strains, in contrast, can enzymatically break the link between melibiose and fructose:

melibiose   fructose

Example of trisaccharide structure

The ale yeasts can then ferment the fructose. However, they cannot ferment melibiose. This property is one of the main features that distinguish the first and second categories of yeast attenuation behavior shown in Table 3.1.

Another trisaccharide of interest is maltotriose, which generally makes up 10% to 15% of the sugars in an all-grain wort. When yeast strains mutate, they often lose their ability to effectively ferment this trisaccharide. In fact, it is now known that only one gene controls the relevant enzymes, and it is easily lost. Thus, yeasts in the third category in Table 3.1 are generally considered dysfunctional.

Temperature. Undoubtedly, the least ambiguous discriminator between brewing yeast strains is their preferred temperature range. The most obvious distinction is between ale and lager strains. The former generally do their best work at ambient temperatures (16–20 °C); however, even in this group, there are individual differences. Scottish strains, for example, prefer the cold side (15–16 °C), whereas most British strains, for example, lose desirable characteristics below 17 °C. Most lager yeast strains do best in the range 8–14 °C; German strains prefer the bottom end of the range at 8–10 °C. Although some special strains (e.g., the so-called steam beer strains)

Table 3.1 **Intrinsic Fermentability of Yeast Strains**		
Category	**Apparent attenuation**	**Comments**
1. Strong attenuators	78%–80%	All fermentable sugars are metabolized. Most lager yeasts (Saccharomyces carls bergensis) fall into this category
2. Medium attenuators	75%–77%	Minor fermentable sugars are not metabolized (see text). Most ale strains (Saccharomyces cerevisiae) fall into this category.
3. Low attenuators	<75%	Minor sugars and some trisaccharides are not metabo lized. In lager strains, this effect is a sure sign of yeast dysfunc tion. There are few ale strains that are intrinsically weak fer menters, but they are not common.

NOTE: A standard pitching (i.e., yeast addition) rate of 15–20 million cells per milliliter of wort is assumed.

can be used at 16 °C, most lager yeasts will start producing unacceptable levels of esters and fusel alcohols above 14 °C.

By-products. Most brewers are sensitive to the characteristic flavor signatures of yeasts and often select strains on that basis. Just about all beers will have one or more esters above their flavor threshold. This statement applies to even delicately flavored beers. For example, Budweiser tends to have a subtle but discernible and attractive apple-like tone, which contrasts with the pineapple-like effect in Coor's products. In fact, these characteristics are the best way to distinguish between the two in blind tasting. These effects are amplified in beers with higher flavor profiles, and hence a strong case can be made that a yeast strain's ester profile is the best criteria to use in selection.

Another by-product of interest is diacetyl. Different strains do show a marked difference in their ability to absorb and reduce diacetyl. Some British ale strains tend to leave diacetyl levels above threshold, and this effect has been used advantageously in select beer styles. On the other hand, most ale brewers prefer diacetyl-reducing strains, since this compound is unstable and can become highly unpleasant under hostile conditions during shipment and storage. Some of the older lager strains, e.g., W-308, have a tendency to leave diacetyl at high levels. Such strains have virtually disappeared from commercial use, but they still are being used to advantage by some homebrewers in special beers. Nevertheless, as a general rule, lager brewers tend to prefer diacetyl reducers and select the strain with the most desirable ester profile for use.

CONTAMINATING YEASTS

Wild or nonbrewing yeasts are probably the most common minor contaminants, particularly with amateur and small-scale commercial brews. These microbes very likely gain access to brewing areas via airborne particles. Indeed, the concentration of yeast cells in air samples taken from brewing areas is about the same as in space devoted to food processing, which in turn is at four and five times the norm. There are four major groups of wild or nonbrewing yeasts that are most relevant to beer: superattenuating yeasts, wine yeasts, special purpose yeasts, and mutant strains.

Superattenuators. Superattenuating yeasts like *Saccharomyces diastaticus* are the easiest to detect since they will ferment a wide spectrum of carbohydrates including dextrins. They also produce very strong phenolic flavors even if present at low levels. One study (see Fix and Fix, 1997) found phenolic undertones and finishing gravities about 10% lower than normal when these yeasts were at a concentration as low as 100 cells per milliliter. Since normal pitching rates run from 10 to 20 million cells per milliliter, this amounts to one contaminating cell per 100,000 to 200,000 culture cells. It is very likely that partially unclean conditions are responsible for these (and more severe) infections.

Wine Yeasts. Amerine et al. (1972, p. 163) noted that "beer yeasts have proven unsatisfactory for the fermentation of crushed grape musts" because beer yeasts impart an undesirable flavor and form too little alcohol. It is interesting that the converse is also true; i.e., in most circumstances, wine yeasts (*Saccharomyes cerevisiae* vars *ellipsoideus*) are undesirable for beer fermentation. They do usually metabolize the same spectrum of carbohydrates as beer yeasts and are much more alcohol tolerant. Unfortunately, their action yields characteristically strong and unpleasant esters that variously recall rotten fruit and/or solvents. Beer and wine yeasts can generally be distinguished on the basis of their morphology. Wine yeasts are smaller and are elliptically shaped. Beer strains tend to be spherical or egg shaped.

Beer and wine do share common contaminating yeasts. *Hansensula* is a wild yeast genus first detected by E. C. Hansen in his fundamental studies on yeast cultures (see Read and Nagodawithana, 1991). Closely related are *Candida* and *Kloeckera*. These yeasts produce films sometimes called "wine flowers" (Amerine and Joslyn, 1973; Amerine et al., 1972). Off-flavors, often recalling oxidative effects, are due to the formation of acetic acid, aldehydes, and esters.

Brewers have long used special yeasts for special beer formulations. The classic example is the use of strains of the *Brettanomyces* genus in lambics (Guinard, 1990). They are powerful ester producers that yield strong and characteristically fruity aromas. In well-aged beer, these effects can achieve the complexity of great wines. The use of these strains is controversial in both beer and wine production, and detractors characterize the "Brett taste" as "medicinal," "mousey," and/or "horsey," to cite but a few descriptors (Fugelsang et al., 1993).

Champagne yeasts have also been used in brewing particularly for strong ales and barley wines. The three most important strains in this multistrain culture are lager yeasts: *S. carlsbergensis, S. uvarum,* and *S. florentinus* (Amerine and Joslyn, 1973). *S. uvarum* was once thought to be identical with *S. carlsbergensis*, but is now regarded as a largely undesirable mutant of the former. *S. florentinus* is alcohol tolerant, which is the raison d'être for using such a mixed (i.e., multi-strain) culture for barley wines. Contaminants sometimes found in some commercially packaged "champagne yeasts" include *S. bayanus* and *S. pastorianus*, to cite but two examples. The effect of the latter is a rather wide spectrum of off-flavors.

Finally, special yeasts like *S. ludwigii* are being used to produce no-alcohol beers. This strain can be metabolize wort sugars, but it will not produce ethanol or other normal fermentation products like esters. This ability tends to be a problem, however, since such beers can be devoid of "beery" characteristics.

When subject to stress, brewing yeasts mutate. They are not as unstable as the typical yeast strains found in basic microbiological research (Reed and Nagodawithana, 1991), but brewing yeasts do have their limits. Adverse environments include excessive alcohol or high carbohydrate concentrations. Thermal stresses are important in a negative sense. Also to be avoided are significant changes in osmotic pressure that comes, for example, when yeasts are suddenly transferred from beer to a different type of medium. Minor mutation is usually reflected in changes in the flocculation and attenuation characteristics discussed above. The most significant types of mutant as far as brewing is concerned are the respiratory-deficient mutants (RDM). This genetic deficiency is reflected first of all in the ability to respire in an oxygen-saturated carbohydrate medium. This ability, in fact, is the basis for most methods used to detect them (Fix and Fix, 1997). More important to brewers are other genetic changes such as the inability to reduce diacetyl and an increased propensity toward fusel alcohol production. The general rule in brewing is that such mutants be kept well below 1% of the pitching yeasts.

On the other hand, brewing strains that have been afforded proper care can and should be repitched if this process can be done in a timely manner after collection. Indeed, there are examples of successful commercial breweries that have reused yeasts collected from previous brews for decades without reculturing (Lieberman, 1980). There is sound scientific theory underlying this practice, which is generally called the birth-scar theory of yeast multiplication (Beran et al., 1966). Attention is given to the condition of the yeast cell wall as a measure of viability because the ability of the yeast cell to have materials move in and out of the cell wall is closely related to viability. The main point of the birth-scar theory is that as a cell reproduces by budding, a birth scar is formed on the mother but not on the daughter. To illustrate this process, consider a mother cell that is reproducing for the first time. The budding process goes as follows:

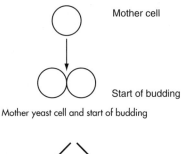

Mother yeast cell and start of budding

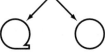

Mother yeast cell with birth scar and daughter without scar

This process is represented in the following diagrams, where the number inside the cell indicates the number of birth scars:

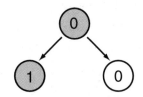

Mother yeast cell (shaded) and daughter cell

This diagram means that a mother cell without birth scars gives rise to two cells, one—the original mother cell—with a birth scar and the other, a daughter cell that is free of scars. The second budding produces the following:

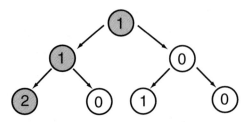

Mother yeast cell (shaded) and three daughter cells

After two divisions, the original mother has two birth scars, the original daughter has one, and there are two new daughters with no birth scars. As this process continues, each individual cell "ages" by the accumulation of birth scars. After a certain number of birth scars have accumulated (usually 10 to 40), the cell wall becomes so damaged that the organism can no longer participate in the fermentation.

The main point is that although individual cells age, the total ensemble of cells does not age. Indeed, since any particular cell gives rise to exactly one cell without birth scars, while picking up a birth scar itself, it follows that at any generation exactly 50% of the cells have no birth scars. Continuation of this logic shows that 25% have one birth scar, 12.5% have two birth scars, etc. One can quickly conclude that well cared-for yeasts can be used indefinitely.

It is felt by some that there may be more to cell aging than birth scars. However, in practical brewing situations departures from the birth scar theory are invariable due to technical errors in brewing. Bacterial contamination and yeast starvation must be avoided. In addition, different yeast strains have different tendencies in regard to mutation, so strain selection is important. Nevertheless, all of these things are well within a brewer's control.

The acquisition of birth scars is not the only way cell walls can become damaged. Fermentation of worts whose gravity exceeds 14 °P can create increased osmotic pressure on the cell wall, a potentially damaging situation (Owades, 1981; Panchal and Stewart, 1980). Transferring yeasts immediately from wort or beer to distilled water results in a dramatic lowering of osmotic pressure, which can also lead to cell wall damage.

FERMENTATION STAGES

The fermentation process can be divided into two stages: the initial period and the Embden-Meyerhof-Parnes (EMP) pathway. This division is somewhat arbitrary because both stages may overlap in time. Nonetheless, the yeast functions are different in the two stages, and so the division is conceptually sound.

The initial period is one of preparation, and it is here that yeast viability is most obvious. Well-fed, vigorous yeast cells will quickly pass through this stage; less hearty yeast cells can have potentially damaging lag times. During this initial period, the yeast cell's activities include cell-wall preparation and oxygen, nitrogen, and sugar uptake. In order to make use of wort nitrogen and sugars, the yeast cell must be permeable. This condition is created by sterol synthesis, for which yeast food reserves, most notably glycogen, are needed. This requirement is why starving yeasts can display erratic behavior at this stage (Quain and Tubb, 1982; Murrary et al., 1984). Oxygen is a crucial nutrient in this process. When the wort has been cooled, it is absolutely essential to add oxygen to it. Alternatively, oxygen can be added to the yeast cells before yeast pitching.

There are industrial fermentations, e.g., those associated with certain synthetic fuels, that avoid use of oxygen. They do this by pitching large quantities of yeasts—typically 100 million cells per milliliter or more—and using media rich in unsaturated fatty acids. These will get the fermentation started and ensure that it will efficiently go to completion. However, it is important to remember that the flavor of the finished product is not an issue with such processes! Indeed, it has been well established (Cahill, 1999) that in beer brewing, overpitching can lead to a broad spectrum of off-flavors, increasing risk of autolysis, clarification problems, and losses of hop flavor and aroma. In addition, unsaturated fatty acids that spill over to the finished beer tend to strongly promote beer staling. (This process is discussed in chapter 4.)

Once the cell walls are prepared, the individual yeast cells start taking in amino acids, elementary peptides, and sugars in a definite order according to the mechanisms described in the next section. Of practical importance, however, are two inhibitory effects created by wort composition and fermentation conditions: maltose inhibition and shock excretion. Worts with large nongrain components (i.e., significant amounts of glucose or fructose) can create a number of problems with the yeast cells' activities. The most important one is the inhibition of the yeast cells' ability to bring maltose through the cell wall. This problem can lead to a long, disordered fermentation. "Shock excretion" refers to the situation where adverse fermentation conditions—most notably high starting gravities and/or high fermentation temperatures—create osmotic-pressure effects on the cell wall. This pressure can cause yeasts to actually reject essential nutrients, most notably wort nitrogen (Owades, 1981; Lewis, 1963; Owades and Jakovac, 1959). Lack of nitrogen will inhibit yeast growth, again resulting in a lengthy and disordered fermentation.

In addition, the initial period is marked by yeast growth (cell division), a buildup of energy reserves, and acidification. All of these activities are of crucial significance for an orderly fermentation. Once the major wort sugars (maltose and maltotriose) are brought into the cell, they are broken down to glucose via enzymatic reactions. The first step in the acidification process is the breakdown of each mole of glucose to 2 mol of pyruvic acid via the following reaction:

Breakdown of 1 mol of glucose to yield 2 mol of pyruvic acid

In this process, 1 mol of the glucose (a six-carbon sugar) is broken down to 2 mol of the pyruvic acid. It is interesting that the formation of pyruvic acid is a branching point for the formation of many by-products like diacetyl. Some of the pyruvic acid forms the product acetyl coenzyme A during the next step:

$$CH_3COCOOH \rightarrow CH_3COS + CoA$$

Acetyl coenzyme A is very important; for one thing, it is the first fermentation product that contains sulfur. This step is another important branch for by-products, most notably esters and fatty acids.

Once the yeast cells have fulfilled their energy and growth requirements, true fermentation begins. The starting point is the pyruvic acid branch and the end point is ethanol, which is a reduction product from acetaldehyde:

Pyruvic acid   Acetaldehyde   Ehtanol

True fermentation process

The overall process starting with the elementary sugars and leading to ethanol is called the Embden-Meyerhof-Parnes or EMP pathway. This process takes each mole of glucose (or fructose) and yields 2 mol each of ethanol and carbon dioxide:

$$C_6H_{12}O_6 \rightarrow 2(C_2H_5OH) + 2(CO_2).$$
$$\text{glucose} \qquad\qquad \text{ethanol}$$

This reaction is sometimes called the classic Gay-Lussac formula after Joseph Louis Gay-Lussac, a nineteenth century French chemist and physicist. Note that the molecular weight of glucose is 180 (i.e., 180 g per mole of glucose) and that of ethanol is 46 (i.e., 46 g per mole of ethanol). Observe that 1 mol of glucose = 2 mol of ethanol, and so

$$\frac{[(180 \text{ g/mol of glucose}) \times (1 \text{ mol of glucose})]}{[(46 \text{ g/mol of ethanol}) \times (2 \text{ mol of ethanol})]} = \frac{180 \text{ g of glucose}}{92 \text{ g of ethanol}} = 1.9565.$$

Therefore, to produce 1 g of ethanol requires 1.9565 g of glucose.

Carl Joseph Napoleon Balling studied the relationship between extract (i.e., wort carbohydrates) (which he expressed as an equivalent amount of sucrose) and ethanol levels from an empirical point of view in order to take into account losses that occur in real fermentations. The main loss is due to the yeast cells' use of carbohydrates for growth as opposed to ethanol production. Balling's measurements showed that

$$2.0665 \text{ g of fermentable extract} \rightarrow 1 \text{ g of ethanol} + 0.9565 \text{ g of } CO_2 + 0.11 \text{ g of losses.}$$

This measurement relates the following:

A = Alcohol content percentage (w/w)
OE = Original extract percentage (w/w); i.e., °P
RE = Residual extract (real extract) percent (w/w); i.e., °P

Indeed, observe that

$$[(2.0665 \times A) + RE] \times 100 \text{ g of extract} \tag{3.1}$$

is required to produce 100 g of beer with an alcohol content equal to A. This is not equal to OE × 100 because of CO_2 production and minor losses. It is in fact equal to

$$(100 + 0.9565 \times A + 0.11 \times A) \times OE = (100 + 1.0665 \times A) \times OE . \tag{3.2}$$

Equating and rearranging equations 3.1 and 3.2 gives

$$OE = \frac{2.0665}{(1 + 0.010665 \times A)} \times A + \frac{1}{(1 + 0.010665 \times A)} \times RE . \tag{3.3}$$

The ASBC table (in *Methods of Analysis,* American Society of Brewing Chemists, 1987) gives a slightly more accurate expression of the form

$$OE = 2A + RE - F,$$

where F is a correction factor given in tabular form. Differences between the two formulas are minor. For example, if $A = 4\%$ (w/w) and $RE = 4.5\ °P$, the correction factor from the ASBC table is $F = 0.255$. This result gives

$$OE = (2 \times 4) + 4.5 - 0.255 = 12.245.$$

On the other hand, equation 3.3 gives

$$OE = \frac{2.0665}{1 + (0.010665 \times 4)} \times 4 + \frac{1}{1 + (0.010665 \times 4)} \times 4.5 = 12.244.$$

Small-scale brewers who do not have the instrumentation to directly measure alcohol levels in beer typically use equation 3.3 to compute A from the original and real extract. For this purpose, equation 3.3 can be rewritten

$$A = \frac{(OE - RE)}{[2.0665 - (0.010665 \times OE)]}. \tag{3.4}$$

Once the fermentation begins, it is important to exclude oxygen because the EMP pathway is an anaerobic process. In fact, if oxygen is introduced during fermentation, the yeast cells will tend to revert to respiration. This process is often called the *Pasteur effect.* (In this context, *respiration* refers to the uptake of O_2 and the release of CO_2 from yeast cells without the production of ethanol.) The reverse of this process—i.e., the inhibition of respiration in favor of fermentation—is called the *Crabtree effect*; it is often seen in conjunction with high-dextrose worts and is usually accompanied by the inhibition of maltose uptake. Most normal brewing strains are affected to varying degrees by the Crabtree effect even in normal beer wort. Super-attenuators such as *Saccharomyces diastaticus* are strongly affected, but *Hansensula* and *Candida* are not (Hough et al., 1981).

Before starting a detailed discussion of the reactions that occur in the fermentations, it is perhaps helpful to make some general points about them. One type of reaction, called *catabolic*, leads to a net increase in metabolic energy, which is stored chemically in the form of ATP (adenosine triphosphate). This high-energy compound is pro-duced from ADP (adenosine diphosphate) during the catabolic reac-tions. The following diagram shows the process as a side reaction, ADP → ATP:

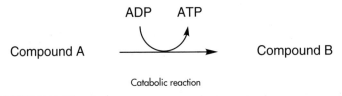

Compound A   Compound B

Catabolic reaction

Anabolic reactions, in contrast, consume energy, as is indicated by the side reaction ATP → ADP. Thus they are the reverse of catabolic reactions, for example:

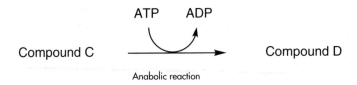

Compound C   Compound D

Anabolic reaction

Redox reactions are also important during the fermentation. There is a duality in redox reactions: One compound is oxidized if and only if another compound is reduced. An important cofactor for the redox reactions in the fermentation is nicotinamide adenine din-ucleotide (NAD). In a typical redox reaction,

---

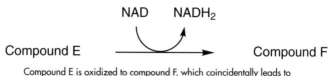

NAD   NADH$_2$

Compound E   Compound F

Compound E is oxidized to compound F, which coincidentally leads to
the reduction of NAD to NADH$_2$

In this redox reaction, a compound (E) is oxidized via the loss of two hydrogen atoms, while NAD is reduced to NADH$_2$ via the gain of two hydrogen atoms. The reverse of this process—in which a compound (G) is reduced while NADH$_2$ is oxidized to NAD—also occurs:

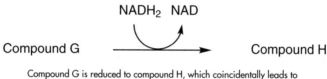

NADH$_2$ NAD

Compound G   Compound H

Compound G is reduced to compound H, which coincidentally leads to
the oxidation of NADH2 to NAD

THE INITIAL PERIOD

The uptake of dissolved oxygen occurs very rapidly, usually within a few hours. The assimilation of wort sugars and nitrogen is slower, the exact rate depending very much on the condition of yeast cell walls. The mechanisms by which yeast wort constituents are transported through the cell membrane are still not understood completely. However, it is known that the sterol content of the cell is a crucial issue; the content must be sufficiently high for the cell wall to have proper permeability (Quain et al., 1981; Aries and Kirsop, 1977).

At the height of the fermentation, when various materials freely pass across the cell membrane, sterol levels are at a peak (about 1% of the total chemical constituents of yeast cells or 10 mg per gram of yeast cells). As the fermentation slows, and particularly as sugar uptake slows, sterol levels drop eventually to about 0.1% or 1 mg per gram of yeast cells. Thus, when yeasts are repitched into fresh wort, a major task for each yeast cell to accomplish is rebuilding the cell's sterol supply through biosynthetic reactions involving cell lipids. One important mechanism is centered around mevalonic acid (a major cell lipid):

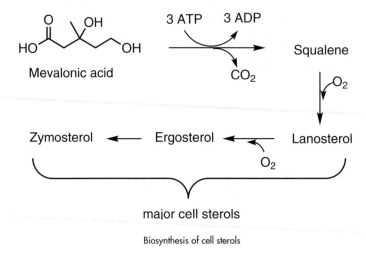

Biosynthesis of cell sterols

Observe that the final step leading to the formation of the major sterols requires oxygen. This is but one reason why poorly aerated cold worts can lead to disordered fermentations. It is interesting that fatty acids derived from wort trub, such as oleic and linoleic acids, can be used as substitutes in the sterol synthesis (Aries and Kirsop, 1977). This requires much less oxygen and allows trub carryover to be used as a substitute for cold-wort oxygenation.

Wort trub can also stimulate yeast metabolism in a purely mechanical manner (Siebert et al., 1986). As the trub particles float in wort, they serve as nucleation sites for CO_2 evolution in the same way that dirt particles suspended in beer cause the evolution of CO_2, which can be seen by bubbles rising from the particles. Saturation of CO_2 in fermenting wort tends to retard yeast metabolism (Arcay-Ledezma and Slaughter, 1984); thus, the presence of the trub particles serves as a yeast stimulant. In spite of these advantages, brewers still prefer clarified worts with minimum trub carryover, if for no other reason than the negative role wort-derived fatty acids play in beer staling.

Observe also that the first part of sterol synthesis is anabolic, meaning that it requires an energy supply for the reactions to take place; three ATP molecules must be used to change each molecule of mevalonic acid to squalene. As a result, in the initial period of fermentation, yeast cells must have internal reserves to supply this metabolic energy. In this regard, two yeast-cell storage carbohydrates are crucial, namely, glycogen ($C_6H_{10}O_5$) and trehalose, a disaccharide ($C_{12}H_{22}O_{11}$) (Quain and Tubb, 1982). Glycogen consists of 1-4 linked chains of 10 to 14 glucose units that are joined by 1-6 links and therefore have a structure similar to malt starch (1-4 and 1-6 refer to the carbon-pair linkage discussed in chapter 1).

The profile for yeast-cell glycogen content throughout a fermentation has the following form:

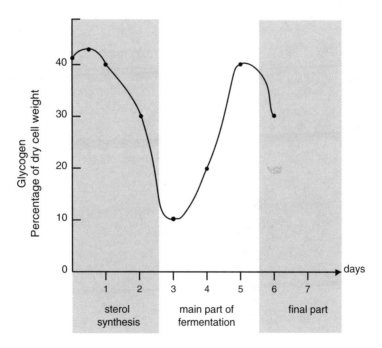

Change in yeast-cell glycogen content over time

Ideally, the glycogen content of pitching yeasts should be at least 40% of the cell's dry weight. The glycogen is rapidly depleted during the synthesis of sterols, which shows how important a role glycogen plays in supplying metabolic energy. Once wort constituents start to enter the cell, the glycogen level begins to increase, until it reaches a maximum and then decreases slightly toward the end of the fermentation. Ideally, yeasts should be repitched 24 to 48 hours after collection, since glycogen reserves are depleted rapidly in storage, (Murrary et al., 1984) although the rate of depletion also depends on yeast strain and brewing conditions. As a general rule, yeasts stored for any length of time should be "fed" with fresh sterile wort to ensure that the storage medium has adequate yeast-assimilable sugars and amino acids.

Since yeast starvation and glycogen depletion are strongly correlated, it is important for brewers to measure the glycogen levels in yeast batches to be pitched. This is particularly true of yeasts that have been stored. There is a simple iodine test for glycogen levels in yeasts (Fix, 1986, 1990).

Once the cell walls become permeable, wort constituents start to enter. The sugar intake is facilitated by enzymes; different enzyme systems are devoted to different sugars. The situation is diagrammed schematically in Figure 3.1.

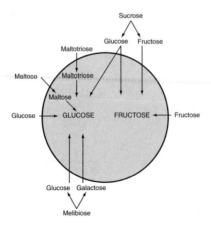

Figure 3.1. Wort sugar intake (Stewart et al., 1983).

The elementary sugars glucose and fructose are generally the first to enter the yeast cell. They are followed by sucrose. Before entering the cell, sucrose is first inverted, or split, into glucose and fructose units by the yeast enzyme invertase. The glucose and fructose units are then taken up by the cell. Yeasts possessing the enzyme melibiase (e.g., most lager yeasts) can utilize melibiose, which is also split outside the cell membrane.

Maltose is brought into the cell at a slower rate than glucose or sucrose. Maltose is transported intact into the cell by the maltose permease enzyme (maltase) and then split inside the cell into two glucose units by the enzyme ß-glucosidase. Maltotriose is generally the last sugar to be used, and its transport mechanisms are similar to those of maltose. However, some of the widely used lager strains, e.g., W-34/70, will take in maltotriose at the same time as it takes in maltose.

The permease enzymes maltase and maltotriase are highly sensitive to wort sugar composition and can be deactivated by high levels of glucose or fructose. This process is called catabolite repression, or sometimes the glucose effect (Siro and Lovgrem, 1980). In an all-grain wort of normal gravity, this effect is not of major importance. However, worts with large amounts of sugar seriously inhibit maltose and maltotriose uptake.

As noted earlier, yeasts take up the various amino acids in a definite sequence. The amino nitrogen (NH_3) is used as a nutrient, while the carbon skeleton (CH_2OH) is sent to the oxoacid pool in the yeast cell for further processing (oxo- means "containing oxygen"). As a general rule, amino acids are not treated separately, but rather in pairs. The Strickland reaction is a major mechanism whereby one amino acid donates hydrogen atoms and the second acid acts as an acceptor. The net result is amino nitrogen and carbon skeletons (oxoacids):

$$R_1 - \underset{\underset{H}{|}}{\overset{\overset{NH_2}{|}}{C}} - COOH + 2H_2O \rightarrow R_1 - COOH + NH_3 + 4H^+$$

H⁺-donor *(continued)*

$$R_2 \overset{\overset{\displaystyle NH_2}{\displaystyle |}}{-} C - COOH + 2H^+ \rightarrow R_2CH_2 - COOH + NH_3$$

H+ - acceptor

Strickland reaction

Some of the amino acids assimilated by yeast cells are excreted back into the wort. Starving yeast cells display a strong propensity for this behavior (to the detriment of the fermentation), and it can also be encouraged by high osmotic pressure created by high wort gravity (Owades, 1981). The overall effect is called "shock excretion" (Panchal and Stewart, 1980; Quain and Tubb, 1982; Murrary et al., 1984; Lewis, 1963).

In normal fermentations, only 50% of the wort amino acids are assimilated by yeasts. Of these, approximately one third are excreted back into wort, although this proportion can be higher under adverse conditions. The situation is diagrammatically represented as follows:

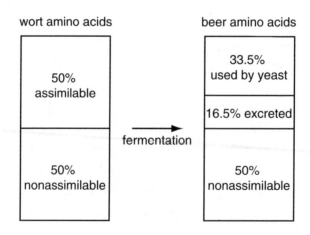

Amino acid usage by yeasts during fermentation

Thus, under normal conditions, 33.5% of the wort amino acids are consumed during the fermentation, while 16.5% of the wort amino acids are assimilated, only to be excreted back into the fermenting wort.

THE EMP PATHWAY
(GLYCOLYSIS)

The first part of the EMP pathway is in effect a precursor to true fermentation in the sense that ethanol is not produced. The redox reaction

$$NAD^+ \rightarrow NADH + H^+$$

plays an important role. Since the NAD is in limited supply, external oxygen is typically needed. It should also be noted that the early parts of the aerobic cycle are energy consuming (anabolic, i.e., side reaction ATP $\rightarrow$ ADP). However, the later parts of the aerobic cycle are catabolic (i.e., side reaction ADP $\rightarrow$ ATP). Various enzyme cofactors that assist enzyme activities—notably Mg^{2+} and Zn^{2+}—are also important. There are five major events in the aerobic cycle, and each one is discussed in sequence. For simplicity, only the fate of glucose units is treated. Fructose has a similar metabolic transformation.

PHOSPHORYLATION

The process of phosphorylation endows glucose with a phosphate structure by expending metabolic energy:

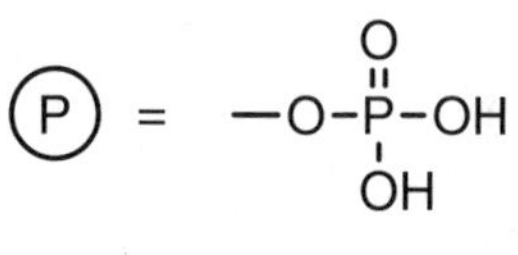

$$\text{(P)} = -O-\overset{\overset{\displaystyle O}{\|}}{\underset{\underset{\displaystyle OH}{|}}{P}}-OH$$

Phosphate structure

The product of phosphorylation is transformed into a fructose analogue, which then gains another P:

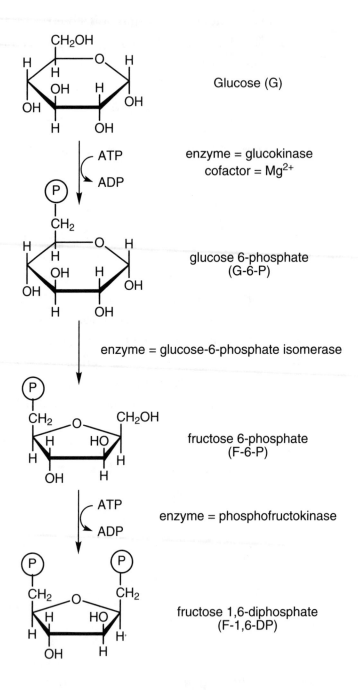

Glucose (G)

enzyme = glucokinase
cofactor = Mg^{2+}

glucose 6-phosphate
(G-6-P)

enzyme = glucose-6-phosphate isomerase

fructose 6-phosphate
(F-6-P)

enzyme = phosphofructokinase

fructose 1,6-diphosphate
(F-1,6-DP)

Glucose phosphorylation yielding two phosphate-bearing fructose units

There is also a minor branch where the glucose 6-phosphate is produced called the hexose monophosphate pathway (HMP). This branch is discussed in the next section.

CARBON SPLITTING

During the second part of the aerobic cycle, the hexose (six-carbon) structure is split into a triose (three-carbon) structure:

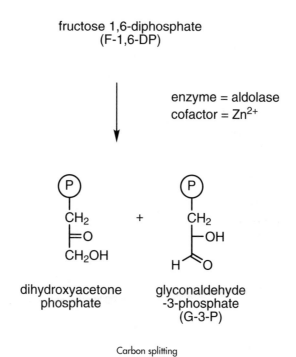

Carbon splitting

This reaction is mediated by the enzyme aldolase and does not require energy, but the presence of zinc (Zn^{2+}, called a cofactor) is crucial. Malt is the primary source of this mineral; thus brewers sometimes add zinc to worts that have large proportions of unmalted grains and/or sugar.

REDOX REACTIONS

At this point in the brewing process, the oxidation-reduction reactions start (NAD is reduced to $NADH_2$), and for the first time, metabolic energy is being produced (ADP to ATP).

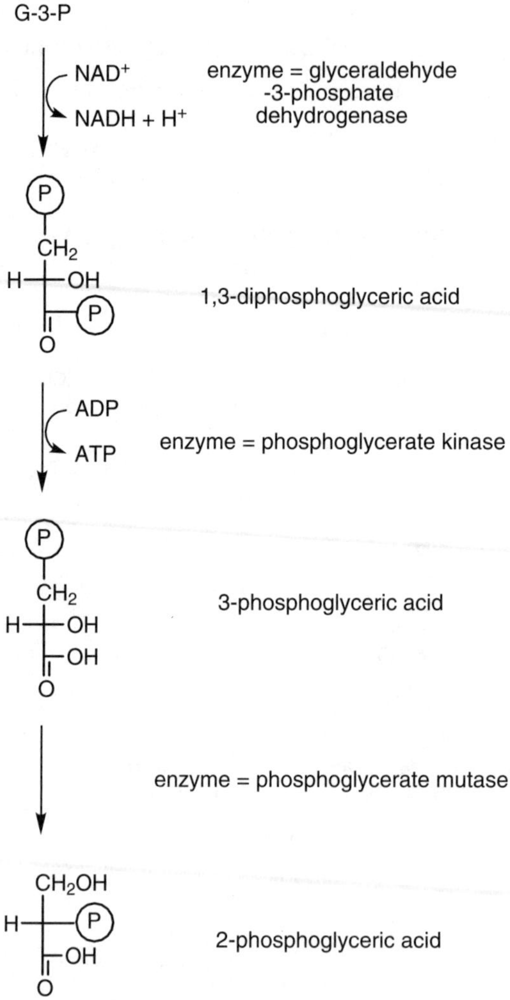
G-3-P
NAD+
NADH + H+
enzyme = glyceraldehyde -3-phosphate dehydrogenase
P
CH2
H
OH
O
1,3-diphosphoglyceric acid
ADP
ATP
enzyme = phosphoglycerate kinase
3-phosphoglyceric acid
enzyme = phosphoglycerate mutase
CH2OH
2-phosphoglyceric acid

NAD reduction to $NADH_2$ and energy storage in ATP

These reactions are mediated by a variety of kinase and dehydrogenase enzymes. The key is maintaining the presence of NAD, which in turn is affected by wort aeration.

FORMATION OF PYRUVIC ACID

Next, pyruvic acid is formed. This process is yet another step in which a cofactor—in this case, magnesium ions (Mg^{2+})—plays a crucial role.

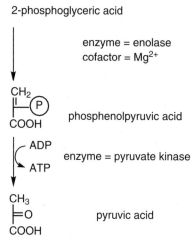

2-phosphoglyceric acid

enzyme = enolase
cofactor = Mg^{2+}

CH_2
$\parallel$ (P)
COOH   phosphenolpyruvic acid

ADP
ATP   enzyme = pyruvate kinase

CH_3
$\models O$
COOH   pyruvic acid

Pyruvic acid formation

FERMENTATION

True fermentation (i.e., production of alcohol) begins when pyruvic acid is formed and terminates with the formation of ethanol.

CH_3
$\models O$
COOH   pyruvic acid

enzyme = pyruvate decarboxylase
cofactor = Mg^{2+}

CO_2

CH_3
$\models O$
H   acetaldehyde

NADH + H^+ enzyme = alcohol dehydrogenase
NAD$^+$ cofactor = Zn^{2+}

Fermentation to produce ethanol

The fermentation step officially ends the EMP pathway. In the next section, alternative branches from this main pathway are discussed.

OTHER METABOLIC PATHWAYS

The diversions off of the main metabolic pathways are, from a microbiological point of view, rather trivial since their products are invariably below 1 mg per liter of beer. This small concentration is in sharp contrast with ethanol, the product of the main (i.e., the EMP) pathway, which is about 40,000–50,000 mg per liter for beers of standard strength. Yet, as far as brewing is concerned, the minor products are of great significance because they have low flavor thresholds. In other words, their flavor influence is far out of proportion to their concentration.

The four most significant minor pathways lead to production of fusel alcohol and esters, diacetyl, phenol, and glycerol. The starting point is at the end of pyruvic acid formation in the EMP pathway. At this point, the metabolically active form of acetate known as acetyl coenzyme A (acetyl-CoA) is formed by the oxidation of pyruvic acid. A key to this process is the compound called coenzyme A (CoA-SH), and this is the first place where a sulfur-containing compound plays an important role. The reaction system is the following:

$$CH_3\text{-}CO\text{-}COOH \xrightarrow[\text{CoA-SH}]{\text{NAD} \quad \text{NADH} + H^+} CH_3CO\text{-}S\text{-}CoA + CO_2$$

pyruvic acid   acetyl-S-CoA

Pyruvic acid to acetyl-S-CoA pathway

This reaction leads to, among other things, pathways that can be used by brewing yeasts to produce fusel alcohols, fatty acids, and esters, and a summary of their production is shown in Figure 3.2.

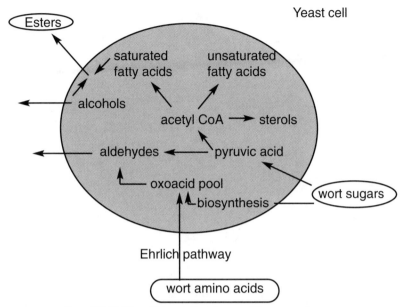

Figure 3.2. Ehrlich and biosynthetic pathways within a yeast cell.

Fusel alcohols are produced from the carbon skeletons of amino acids that were derived either from wort (Ehrlich mechanism) or from sugar metabolism (biosynthetic mechanism). The carbon skeletons wind up in a collection of oxoacids from which aldehydes are formed. Later, the aldehydes are transformed into the corresponding fusel alcohol. This mechanism can be represented as seen on page 108). For example, if the amino acid is leucine, then R is

$$R = \begin{matrix} H_3C \\ \diagdown \\ / \\ H_3C \end{matrix} - CH_2$$

Structure of leucine

and the final product is isoamyl alcohol.

The previous example describes the production of volatile aliphatic alcohols. Nonvolatile aromatic alcohols are produced in a similar way from phenolic amino acids:

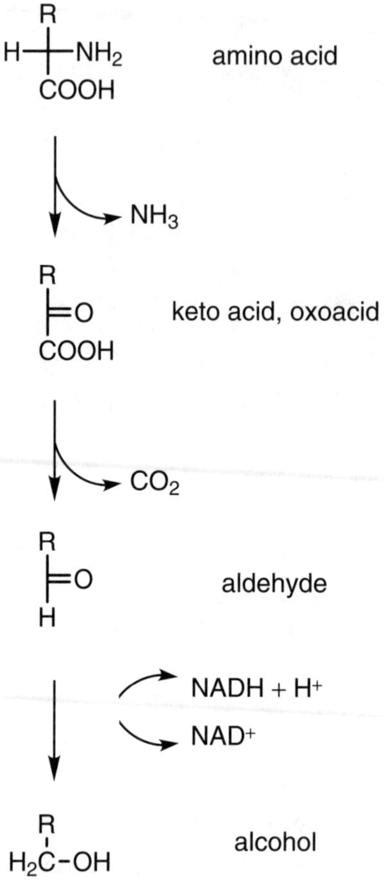
R
H–NH2
COOH
amino acid
NH3
R
=O
COOH
keto acid, oxoacid
CO2
R
=O
H
aldehyde
NADH + H+
NAD+
R
H2C-OH
alcohol

Production of fusel alcohol

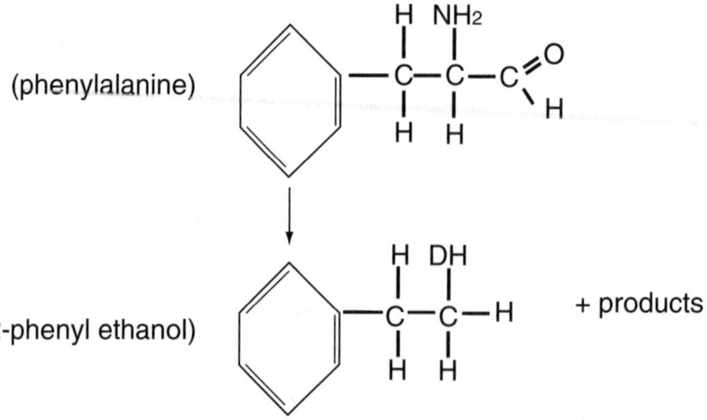
(phenylalanine)
H NH2
C C C
O
H
H H
(2-phenyl ethanol)
H DH
C C H
H H
+ products

Production of phenyl alcohol from phenylalanine

Tyrosol is another example of a nonvolatile aromatic alcohol that can be formed by this process.

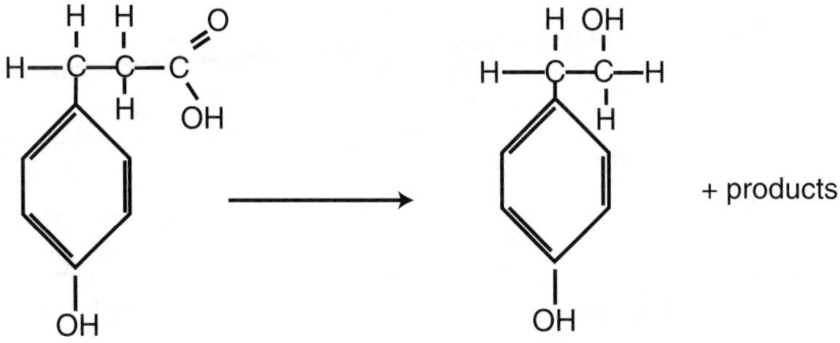

Production of tyrosol from tyrosine

As noted above alcohols have structures of the form

$$ROH$$

Alcohol structure

For example, ethanol has the structure

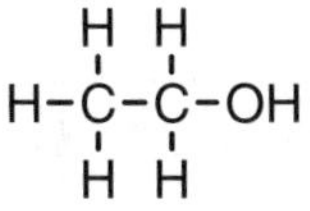

Ethanol

So, for ethanol, $R = C_2H_5$. The major fusel alcohols relevant to beer are displayed in Table 3.2.

The acetyl-CoA also serves as a branch for fatty-acid synthesis and production; this branch is used more or less by certain yeast strains that have particular synthesizing abilities (Narziss et al., 1984). Of greatest interest are the saturated fatty acids displayed in Table 3.3. These are to be distinguished from the metabolism of unsaturated fatty acids involved in sterol synthesis that was discussed previously in

Table 3.2 — Fusel Alcohols

Name	Structure	Range (mg/L)	Threshold (mg/L)	Flavor
Propanol (propyl alcohol)	$H-\overset{H}{\underset{H}{C}}-\overset{H}{\underset{H}{C}}-\overset{H}{\underset{H}{C}}-OH$	10–40	600–800	Alcohol—rough aftertaste
Butanol (butyl alcohol)	$H-\overset{H}{\underset{H}{C}}-\overset{H}{\underset{H}{C}}-\overset{H}{\underset{H}{C}}-\overset{H}{\underset{H}{C}}-OH$	5–60	160–200	Alcohol—rough aftertaste
Isobutanol (2-methyl-1-propanol)	$H-\overset{H}{\underset{CH_3}{C}}-\overset{H}{\underset{H}{C}}-\overset{H}{\underset{H}{C}}-OH$	10–60	180–200	Alcohol—rough aftertaste
Active amyl alcohol	$H-\overset{H}{\underset{H}{C}}-\overset{H}{\underset{H}{C}}-\overset{H}{\underset{CH_3}{C}}-\overset{H}{\underset{H}{C}}-OH$	100–110	40–130	Alcohol—banana solvent
Isoamyl alcohol	$H-\overset{H}{\underset{H}{C}}-\overset{H}{\underset{CH_3}{C}}-\overset{H}{\underset{H}{C}}-\overset{H}{\underset{H}{C}}-OH$	100–110	40–130	Alcohol—banana solvent
Phenolethyl alcohol	(phenyl ring)$-\overset{H}{\underset{H}{C}}-\overset{H}{\underset{H}{C}}-OH$	100–200	10–80	Roses—bitter, chemical, medicinal
Tyrosol	$HO-$(phenyl ring)$-\overset{H}{\underset{H}{C}}-\overset{H}{\underset{H}{C}}-H$	100–200	10–80	Roses—bitter, chemical, medicinal

Table 3.3 — Saturated Fatty Acids from Yeast Metabolism

Compound	Formula	Threshold (mg/L)	Flavor
Caproic acid (hexanoic: six carbons)	$CH_3(CH_2)_4COOH$	8	goaty sweaty fatty
Caprylic acid (octanoic: eight carbons)	$CH_3(CH_2)_6COOH$	13–15	goaty fatty
Capric acid (decanoic: ten carbons)	$CH_3(CH_2)_8COOH$	10	soapy

this chapter. The compounds cited in Table 3.3 should also be distinguished from the fatty acids associated with hop staling (e.g., valeric and butyric acids discussed in chapter 4).

Once alcohols are present, there is the possibility of their combining with organic acids to form esters. In this process, an alcohol R_1CH_2OH and an acid R_2COOH react as follows:

$$R_1 - OH + R_2 - \overset{\displaystyle O}{\underset{\displaystyle OH}{\overset{\|}{C}}} \longleftrightarrow R_1 - O - \overset{\displaystyle O}{\overset{\|}{C}} - R_2 + H_2O$$

(alcohol)   (carboxylic acid)

Ester production

The end products are water and an ester compound. Observe that the ester compound contains the side chains R_1 and R_2 of both the alcohol and the organic acid that participated in the reaction. Esters are usually named to reflect both the alcohol and the organic acid. For example, ethyl alcohol and acetic acid react to produce ethyl acetate. Over 90 esters are believed to be present in beer, although the three listed in Table 3.4 are the most important: ethyl acetate, isoamyl acetate, ethyl hexanoate (ethyl caproate). Concentrations of ethyl acetate are generally near their threshold and fall into the range of 15 to 20 mg/L, but levels in certain English ales have sometimes been reported in the 30 to 40 mg/L range (Piendl, 1970–1990). This ester usually produces fruity tones, but because it also is associated with a solvent-like undertone, most brewers generally try to keep it in the range 15 to 23 mg/L.

Another flavor-active ester is isoamyl acetate, which is formed from isoamyl alcohol and acetic acid. Isoamyl alcohol is a minor alcohol in beer, but the associated ester has a very strong aroma and is an important flavor component in beer. It is called a "banana ester" because of its sensory characteristics. Its threshold is 3 mg/L and in

Table 3.4	**Principal Esters**		
Compound	Formula	Threshold (mg/L)	Flavor
Ethyl acetate	$CH_3COOC_2H_5$	33	Fruity with solvent undertone
Isoamyl acetate	$CH_3COO(CH_2)_2CH(CH_3)_2$	3	Bananas
Ethyl hexanoate (ethyl caproate)	$C_5H_{11}COOC_2H_5$	123	Apples

most lagers is found in the range 1.2 to 2.5 mg/L (Piendl, 1970–1990). Levels near 6 mg/L have been reported in certain English ales, where it is a valued primary component of the flavor profile.

The final ester in Table 3.4 is the so-called "apple ester," ethyl hexanoate. This ester has a very strong flavor reminiscent of apples, which may or may not be attractive, depending on the beer style.

In most beers the perception of estery notes involves a contribution from many different esters. This results in many different flavor (i.e., taste and smell) sensations.

It is important to emphasize that ester production is not a spontaneous reaction. Merely combining a fatty-acid material (like soap) with an alcohol solution generally will not cause reactions to take place. A catalyst is needed, and in brewing, the catalysts are enzymes in brewing yeasts. Thus, ester production, like fusel alcohol production, is an important part of the signature of a yeast strain. Different strains display entirely different propensities to produce esters (Narziss et al., 1984).

Brewing procedures are also quite important to ester formation. These are highly varied and sometimes seemingly contradictory. For example, very high rates of yeast growth tends to encourage activities at the acetyl-CoA branch and hence to encourage ester formation. Very high wort O_2 levels tend to induce rapid yeast growth and therefore to boost ester formation. On the other hand, it is well documented that poorly oxygenated wort also tends to encourage ester production (Kieninger, 1977). A less ambiguous example involves the fermentation of high-gravity wort. It is not surprising that the higher the concentration of fermentable sugars, the greater probability that the reactions displayed in Figure 3.2 will occur. A consequence that was first discovered with the advent of high-gravity brewing is that the gravity effect is superlinear. For example, everything else being equal, producing an 8 °P beer by fermenting a 16 °P wort and the diluting by a factor of two will lead to ester levels that are significantly higher than the fermentation of a 8 °P wort (Kieninger, 1977).

Pathways leading to the formation of ketones have been of great interest to brewers throughout the twentieth century. Of the various ketone compounds, diacetyl (see Table 3.5) is the most important, and the term *sarcina sickness* has historically been used to describe this defect in beer.

In his fundamental studies, Pasteur showed that microorganisms, which are now called gram-positive lactic acid bacteria, can produce

Table 3.5	**Important Ketones**		
Name	**Structure**	**Threshold (mg/L)**	**Flavor**
diacetyl	$H_3C-C-C-CH_3$ (O O)	0.1–0.15	buttery
α-acetolactic acid	$H_3C-C-C-COOH$ (O OH, CH_3)	–	–
acetoin	$H_3C-C-C-CH_3$ (O H, OH)	1.0	fruity musty musty
2,3-butanediol	$H_3C-C-C-CH_3$ (OH OH, H H)	–	neutral
2,3-pentanedione	$H_3C-C-C-C_2H_5$ (O O)	1.0	honey
α-acetohydroxy-butyric acid	$H_3C-C-C-C_2H_5$ (O COOH, OH)	1.0	rubber

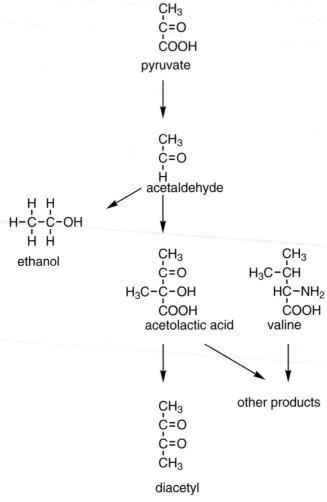

Figure 3.3. Pathway of diacetyl formation.

these disorders in beer (Pasteur, 1876). However, the compound was not identified in Pasteur's work since there was no way at that time of isolating or measuring it. This deficit remained true until the mid-1900s, although there was a growing suspicion that diacetyl-like elements were important. (See the example in the brewing reference manual by A. L. Nugy [1937] that was widely used during the 1930s and 1940s.)

Breakthroughs that occurred in the late 1950s and early 1960s were in many respects almost as revolutionary as Pasteur's work

(Wainwright, 1973). A major discovery was that normal brewing strains were also involved in diacetyl production, and under favorable conditions, they can also reduce diacetyl. The diacetyl-forming mechanism that was discovered starts with pyruvic acid and acetaldehyde, which are transformed to α-acetolactic acid inside the yeast cell. This compound leaks outside the cell, where apparently it is spontaneously oxidized to diacetyl (Wainwright, 1973; Morrison and Bendrak, 1987; Nakatani et al., 1984). This process is schematically represented in Figure 3.3.

This process is inhibited by the synthesis of the valine, an amino acid that enters the cell in the middle of the fermentation. The inhibitory effect arises in part from the fact that valine synthesis utilizes α-acetolactic acid and thus reduces the amount available for diacetyl formation. In addition, α-acetolactic acid can be transformed to valine inside the cell, as shown in Figure 3.4. This possibility means that diacetyl and valine levels are correlated. Under normal conditions, the fermentation curves for both have the following form:

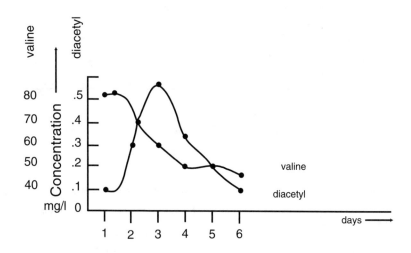

Change in concentrations of diacetyl and valine over time.

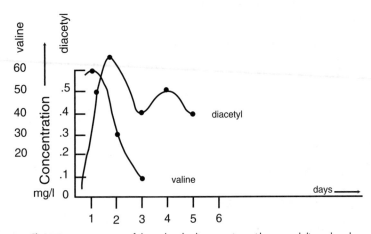

Figure 3.4. Pathway of transforming α-acetolactic acid to valine.

Hence, as valine is assimilated and synthesized (see Fig. 3.3), diacetyl formation is repressed, and the diacetyl reduction mechanisms described next dominate. However, if the fermenting wort is depleted of valine, a second diacetyl peak can emerge.

Change in concentrations of diacetyl and valine over time with a second diacetyl peak

Normally, worts with high malt concentrations will have adequate valine levels, although second diacetyl peaks can build in worts with low malt levels.

The breakdown of α-acetolactic acid to diacetyl occurs outside the yeast cell. In fact, residual α-acetolactic acid that spills over from the yeast cells into beer can spontaneously oxidize to diacetyl. Also, brewing yeasts have enzyme systems capable of reducing diacetyl to less offensive compounds (see Figure 3.5).

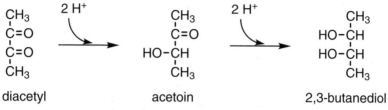

diacetyl   acetoin   2,3-butanediol

Figure 3.5. Pathway of enzymatic reduction of diacetyl by yeast cells.

It must be emphasized that the reduction mechanism requires that the yeast cell reabsorb the diacetyl that is formed. In this regard, major differences are seen between yeast strains, as is illustrated in Table 3.6 for three lager strains.

Table 3.6	**Diacetyl Formation in Three Yeast Strains**		
Day	**Diacetyl level (mg/L)**		
	W-206	W-34/70	W-308
1	0.18	0.15	0.18
3	0.25	0.20	0.48
5	0.23	0.18	0.92
7	0.18	0.14	0.75
9	0.14	0.09	0.65

Brewing conditions are also important. The following problems are the most typical:

- Unclean conditions causing diacetyl production via bacteria
- Excess oxygen causing diacetyl production via redox reactions
- Improper wort composition producing deficiencies in amino acids like valine in low-malt-content worts
- Defective yeast cells—mutants of brewing yeasts typically are good diacetyl producers but poor diacetyl reducers

As a consequence, brewers worldwide now regard diacetyl level as a quality-control parameter quite apart from whatever flavors they may produce in beer. Thus, in the period prior to the 1950s, diacetyl levels in U.S. beers were typically found in the range of 0.2 to 0.3 mg/L, but diacetyl levels today are rarely above 0.05 mg/L (Fix, 1993).

However, some regional breweries in the United States and elsewhere have kept their beers at the pre-1950 diacetyl levels. In fresh beers, many consumers actually find diacetyl at these levels to be attractive. The vanilla tone, which is often confused with caramel flavoring, definitely adds to the smoothness of beer. The problem is that diacetyl at these levels is unstable and eventually will turn into unpleasant flavor notes. This is particularly true of bottled beer with headspace air. As a consequence, these beers typically have a short shelf life and do not travel well.

The other ketone of interest is 2,3-pentanedione, which together with diacetyl forms a beer's vicinal diketones (VDK; aromatic volatile compounds). The flavoring threshold of 2,3-pentanedione is a tenth the strength of diacetyl, and it is rarely an important beer constituent. However, there are well-known Belgian ales that fully display the "honey-flavor" characteristics associated with these ketones.

Certain brewing strains display a propensity for pathways leading to phenolic products that have rather low flavor thresholds. The formation of phenol alcohols cited previously is an example. The other phenol of interest is 4-vinyl guaiacol. It has a medicinal flavor tone and a flavor threshold of about 0.3 mg/l. The following is a typical metabolic reaction:

Pathway to form 4-vinyl guaiacol

None of the major lager strains will metabolize ferulic acid, a minor wort product, and this statement is also true of most ale strains, although there is considerable variation among different strains. Perhaps the extreme case is the strain of top-fermenting yeasts used to produce Bavarian wheat beers, where 4-vinyl guaiacol levels reach the 0.3 to 0.6 mg/L range. Here one finds a "clovelike" phenol flavor that is regarded as a normal characteristic of this beer style.

As already noted, if oxygen is introduced during fermentation, yeasts will tend to revert to aerobic growth via respiration via the process called the Pasteur effect. The glycerol effect is yet another process that arises because of the inhibition of respiration. If the oxidized product NAD is in short supply, then metabolic activities starting during the redox reactions in the EMP pathway will take a different turn. This change often happens after hexose-splitting dihydroxyacetone phosphate is transformed to glycerol via the mechanism shown in Figure 3.6.

The glycerol so produced usually carries over into the finished beer. Observe that this process increases the supply of NAD. This is a minor process in well-oxygenated worts of normal gravity; however, in high gravity and/or weakly oxygenated worts, glycerol levels in finished beer can exceed flavor thresholds and are rarely seen as favorable.

BACTERIA AND BACTERIAL METABOLISM

Single-cell microorganisms are ubiquitous in nature, including all brewing environments. The main reasons why beer has been brewed for so many centuries is that it is safe. Pathogenic bacteria, which are so dangerous in fermented foods, simply cannot survive in beer because of the low pH (3.8 to 4.5) and the alcohol content. In fact, a case can be made that the small number of bacteria that cause flavor problems in beer actually increase its nutritive value. For example, the version of brewer's yeast sold at health food stores contains nontrivial levels of gram-negative rods and cocci. On the other hand, although low levels of bacteria in beer do not cause health problems, they can reduce

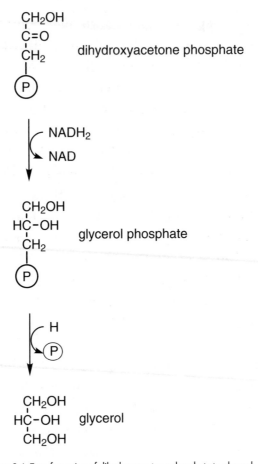

CH_2OH
|
C=O
|
CH_2   dihydroxyacetone phosphate
|
(P)

NADH_2
NAD

CH_2OH
|
HC-OH
|
CH_2   glycerol phosphate
|
(P)

H
(P)

CH_2OH
|
HC-OH   glycerol
|
CH_2OH

Figure 3.6. Transformation of dihydroxyacetone phosphate to glycerol.

what are potentially great beers to mediocrity. For example, in the majority of cases I have seen where a beer lacks an attractive "malty character," the culprit has not been malt types or mashing schedules. Rather it has been the presence of bacteria at levels that are not high enough to destroy the beer, yet are high enough to contribute a thinning and/or drying finish that greatly interferes with malt flavor.

This book takes a functional approach to the classification of bacteria relevant to beer, much as was done with yeasts. Therefore, the following are the three major categories that are covered: the lactic acid group, the acetic acid group, and the sulfur-producing bacteria.

LACTIC ACID BACTERIA

The lactic acid group consists of two genera, *Lactobacillus* and *Pediococcus*. The former genus comprises gram-positive, rod-shaped bacteria ranging up to 1 μm in diameter and 5 to 120 μm in length. Propagation of these microbes under laboratory condition requires strict anaerobic conditions; however, strains isolated in brewery-contamination studies are generally facultative anaerobes (i.e., they display a slight tolerance to oxygen). They are relatively insensitive to hops and tolerate ethanol up to 5% (w/w).

Brewing *Lactobacillus* bacteria are sometimes called heterofermentative since they can produce many major products including lactic acid, ethanol, and CO_2. Their minor products—namely, diacetyl and 2,3-butanediol—are just as important from the point of view of beer flavors. The metabolic pathways used by these microbes are different from the EMP pathway because lactobacilli lack key enzymes (notably aldolase and hexose isomerase). The pathways are displayed in Figure 3.7.

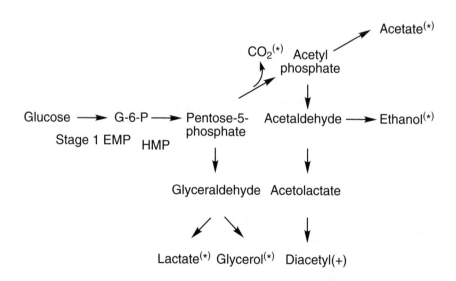

Figure 3.7. The HMP (hexose monophosphate) and phosphoketolase pathways for Lactobacillus. (*) = major product, (+) = minor product.

There are special strains of lactobacilli that essentially have only lactic acid as a product. These have played a crucial role in the production of sour beers, e.g., Belgian and Berliner White beers, as well as indigenous styles like Kentucky Common Beer (Zimmermann, 1904; Wahl, 1908). These strains are also used in mash acidification, as was discussed in chapter 1. As noted there, these strains are rare, and brewers who have had success with them invariably culture the bacteria like yeasts. An important example of a lactic acid producer is *Lactobacillus delbruckii*. It also produces small amounts of other acids (notably acetic, propiolic, formic, buytric, and pyruvic acids), but at levels that are relatively unimportant to beer flavor (Scheer, 1988).

The genus *Pediococcus* is the only cocci bacteria known to grow in beer. They appear in pairs or tetrads and are 0.8 to 1.0 μm in diameter. These microbes are usually called homofermative since they have lactic acid as their only major product. This statement, however, is somewhat misleading since they also produce obnoxious levels of "minor products," most notably diacetyl. They follow the EMP pathway up to the formation of pyruvic acid. Instead of then producing ethanol, these microbes reduce pyruvic acid to lactic acid:

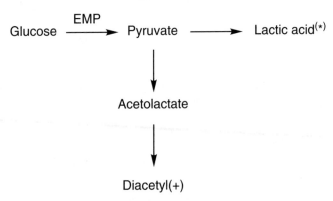

Figure 3.8. Pathway for Pediococcus. (*) = major product, (+) = minor product.

ACETIC ACID BACTERIA

Acetic acid bacteria are gram-negative aerobes that are ubiquitous in all breweries (Ingledew, 1979). This group includes *Acetobacter* and

Acetomonas. They are strongly oxidizing and convert ethanol to acetic acid. They are alcohol and hop tolerant. Their most common occurrence is in partially filled storage tanks, but they also occur in partially filled draught beer containers into which air has leaked (e.g., beer that has been on tap for extended periods).

Sulfur-producing bacteria are gram-negative microbes that consist of many different genera that occur in all stages of brewing. The coliform group are the classic wort spoilers. They are aerobic or facultatively anaerobic rods that are highly hop, ethanol, and pH sensitive. As a consequence, they die off early in the fermentation; however, if they are present at too high of a level in wort, they can create products that are retained and end up in finished beer. These include DMS and H_2S, which have characteristic tones like rotten vegetables and/or overcooked corn.

Closely related is a gram-negative microbe called *Obesumbacterium proteus.* This genus is also ethanol, hop, and pH sensitive. These microbes are short, fat rods (1.5 to 4 µm in length and 0.8 to 1.2 µm in width) and are normally found in pitching yeasts. They form various products in the early part of the fermentation. Table 3.7 gives an example (taken from Ingledew, 1979) for contaminated yeasts. Owing to the low flavor threshold of DMS, the concentrations reported in Table 3.7 would be obnoxious and unacceptable in normal beer.

Perhaps, the most feared gram-negative bacteria are *Megasphaeria* and *Pectinatus* (Lee et al., 1980). These are anaerobic rod-shaped

Table 3.7	**Metabolic Products of** *O. proteus*
	Amount of carbohydrates used (%)
Ethanol	8.5
Lactate	1.5
DMS	3

bacteria that are 2 to 35 μm in length and 0.7 to 0.8 μm in width. Unlike the previously discussed bacteria, these genera are ethanol, hop, and pH tolerant. As a consequence, they are typically found near the end of the brewing cycle. An increase in DMS during maturation is a sure sign of their presence.

A closely related microbe is *Zymomonas*. It is a plump anaerobic rod 2 to 3 μm in length and 1 to 1.5 μm in diameter. It is pH, hop, and ethanol tolerant. *Zymomonas* has a strong preference for glucose and will create unpleasant levels of acetaldehyde and H_2S in a matter of hours. This bacteria genus has been a traditional problem for brewers of cast-conditioned ales and bottle-conditioned beers, especially in warm weather. It has also been reported in breweries under excavation. Other than these occurrences, this bacteria is rarely found in normal brewery environments. It is for this reason that *Megasphaeria* and *Pectinatus* may in fact be mutant *Zymomanas* that have learned to adapt to a standard brewery environment.

Most of the bacteria in the sulfur-producing group follow the Entner-Doudoroff (ED) pathway. It is similar to the EMP pathway in that one unit of glucose gives two units each of ethanol and CO_2. The mechanisms are different, however, and in the bacterial cycle, unpleasant sulfury products such as H_2S and DMS are invariably created as well. The overall outline of the E-D pathway is given in Figure 3.9.

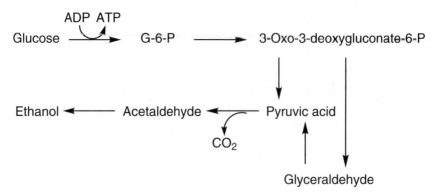

Figure 3.9. The E-D pathway.

In the course of the electron transfers of the E-D pathway, the reduction of DMSO to DMS is encouraged. As noted in chapter 1, it was once thought that DMS could be created in the fermentation by yeast-induced reduction of DMSO. In the absence of bacteria, this process occurs in amounts that are not relevant to the finished beer flavor. However, the extent to which DMS will be "scrubbed out" in the fermentation will vary with different yeast strains and different fermentation conditions. This effect is illustrated in Figure 3.10.

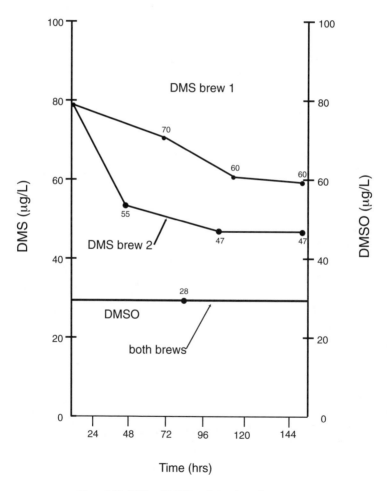

Figure 3.10. DMS and DMSO evolution during fermentation.

Oxidation

4

A strong case can be made that staling is the most common defect in packaged beer found in the trade. In some sense, this predicament is to be expected with any natural organic food like beer.

Additives to cover the effects of staling have been used in the past, but since about 1990, there has been a declining interest in them. Potassium metabisulfite ($K_2S_2O_5$) is an example. Although it still finds favor in wine making and various areas of food processing, the use of potassium metabisulfite has all but disappeared in brewing. Bisulfites work by binding with staling aldehydes to mask the presence of the latter. Unfortunately, these bonds are rather short-lived in beer, and when they are broken, the staling aldehydes fully reveal their presence. Worse still, the bisulfite component can undergo other reactions that produce unpleasant H_2S and/or mercaptan notes.

Another type of additive includes the antioxidants that are neutral to beer flavor. Ascorbic acid (vitamin C) is an example, but unfortunately, it and analogous compounds are not very effective.

Therefore, it is obvious why "nonadditive brewing" is gaining favor. The effective additives tend to have unacceptable side effects, and the neutral additives rarely seem to work. As a consequence, the best course for brewers is to master the fundamental mechanisms that lead to staling and then to use this knowledge to optimize equipment and procedures in order to achieve a respectable, albeit finite, shelf life. Methods for such optimization are covered in, for example, in the references by Bamforth (1999) and Fix (1998). Following C. D. Dalgliesch (1977), it is useful to characterize staling in terms of three basic stages as shown in Figure 4.1.

- Stage A is the period of stable, "brewery-fresh" flavor.

- Stage B is a transition period in which a multitude of new flavor sensations can be detected.
- Stage C products are the classic flavor tones involved in beer staling.

Stage A beer is pristine in flavor. During stage B, Dalgliesch described a decline in hop aroma, a decline in hop bitterness, an increase in "ribes aroma" (or sometimes "catty" flavor), and an increase in sweet, toffee-like, or caramel tones. The terms *ribes* (or *currant*) and *catty* are widely used in the United Kingdom and Scandinavia to recall overripe or spoiled fruit or vegetables. Some tasters cite a "black currant" tone (Hardwick, 1978). In truth, these terms describe a wide spectrum of negative flavors developed when beer is in stage B. Toffee or caramel flavors can come from many sources, but those associated with staling will invariably have unattractive cloying notes. These effects are enhanced by residual diacetyl and also by excess heat treatment of wort. Finally, stage C products range from papery or leathery to sherry- or vinegar-like notes.

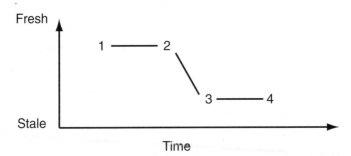

Figure 4.1. The stages of beer flavor evolution.

Although it is not treated in the work by Dalgliesch (1977) or Hardwick (1978), a beer stage D also exists in which stage C flavors have evolved into a kaleidoscope of flavors that in very special formulations—Rodenbach's Grand Cru comes to mind—recall the subtlety and complexity of great wines. It must be emphasized that this process takes years, not months, of maturation.

Barry Axcell and Phil Torline, in an interesting if provocative article, argued that most beers are consumed during stage B (Axcell and Torline, 1998). During this period, beer flavors undergo discernible changes, and the authors suggested that these changes are at the root of consumer dissatisfaction. Among other things, they cited the so-called import paradox as partial evidence of this theory, the paradox being that a definite proportion of the beer-consuming population actually prefers beers in stage C. (Here "import" means any beer consumed at a significant distance from where it is brewed.) These authors noted both the stability of flavor in stage C and "learned prejudices" (such as prestige of the beer and packaging) as the keys to this paradox.

M.C.Meilgaard, for one, has been sharply critical of stage C flavors because they are one-dimensional (Meilgaard, 1991). He stated, "I think it ranks as an all-American scandal that fully half of all the interesting and unusual packaged beers that are on the market get to us so oxidized that staleness and cardboard are the main flavor tones." This is why most successful brewers try to produce beers in which stage A flavors are as stable as possible for as long as possible. If they, or their consumers, prefer stage B or C (or D for that matter), then putting the beer aside for the required time is all they need to do. Staling can go in only one direction!

It is useful to make a distinction between hot-side aeration and cold-side aeration. The *hot side* includes events from mashing through wort chilling. When the wort is cool, oxygen is added to the wort or, more commonly these days, to the yeasts, before the yeasts are pitched. At this point in the process, oxygen serves as an valuable yeast nutrient. Once the fermentation has started, however, oxygen returns to being a negative element. At this point, what is referred to as the *cold side* starts and continues until the beer is consumed.

As a general rule, cold-side aeration involves auto-oxidative processes, i.e., processes in which staling products are produced by the direct attack of molecular oxygen on beer constituents, including aldehydes, hop constituents, phenolic compounds, and, most important, ethanol and other alcohols. Cold-side aeration can arise during the processing of fermented beer; air pickup during filtration causes a particular concern. Thus, the most critical time associated with cold-side aeration is during the packaging of beer. The negative effect of high air levels has been well documented (see, e.g., Fix, 1998). However, it cannot be overemphasized that thermal and mechanical abuse during the storage of packaged beer can greatly accelerate staling even with low headspace air levels, as shown by Fix and Fix (1997).

Hot-side aeration is fundamentally different because of the speed of redox reactions that occur at elevated temperatures. For example, introducing 1 mL of oxygen per liter of wort at 70 °C will start reactions that consume the free oxygen in seconds. The oxidized wort constituents will ultimately cause flavor problems in packaged beer via the chemical mechanisms described in the next section. Doing the same for wort at 20 °C will have virtually no effect. The free oxygen will remain an inert gas for days and, in practical brewing situations, will either be consumed by yeasts or removed by CO_2 evolution during fermentation before it can oxidize wort constituents.

The most deleterious effect of hot-side aeration is associated with long-chain staling aldehydes like trans-2-nonenal. Because of the practical importance of these compounds, an entire section later in this chapter is devoted to them.

BASIC MECHANISMS

This section considers two basic mechanisms that are major factors in beer staling. The first is auto-oxidation in which molecular oxygen directly acts on a compound to produce an oxidized product:

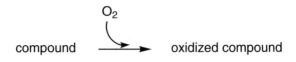

Auto-oxidation reaction

The second mechanism is more subtle, and it is sometimes called *oxidation without molecular oxygen*. It involves electron transfer during which an oxidized compound is reduced in conjunction with a reduced compound becoming oxidized.

The simplest example of auto-oxidation is the oxidation of ethanol to acetaldehyde via

$$\text{2 CH}_3\text{CH}_2\text{OH} \xrightarrow{\text{O}_2} \text{2 CH}_3\text{CHO} + \text{2 H}_2\text{O}$$

       ethanol   acetaldehyde

Oxidation of ethanol to acetaldehyde

This reaction reverses the reduction step at the end of the EMP pathway discussed in chapter 3. Other alcohols, e.g., the fusel alcohols also discussed in chapter 3, can be oxidized by a similar mechanism:

$$\text{2 RCH}_2\text{OH} \xrightarrow{\text{O}_2} \text{2 RCHO} + \text{2 H}_2\text{O}$$

       alcohol   aldehyde

Oxidation of other alcohols to yield an aldehyde

The reappearance of these aldehydes invariably leads to unpleasant flavors (Fix, 1998). In addition, the aldehydes can further oxidize to an acid via

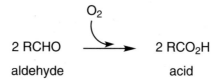

$$\text{2 RCHO} \xrightarrow{\text{O}_2} \text{2 RCO}_2\text{H}$$

       aldehyde   acid

Oxidation leading to acid formation

The so-called vinegar process involves acetaldehyde and acetic acid whereby

$$R = H_3C .$$

Such reactions generally occur on the cold side of the brewing process, i.e., they are an important part of flavor disorders due to cold-side aeration.

The case of iso-α-acids is very important because beer staling invariably involves transformations of hop flavors and these transformations tend to be a signal that staling is underway. The most common mechanism involving iso-α-acids removes or cleaves the acyl side chain R

Acyl side chain R in an iso-α-acid

from the iso-α-acid. This reaction leads to the formation of fatty acids having unmistakable stale and cheesy tones. For isohumulone, whose structure is shown in chapter 2, the structure of the side chain R is as follows:

Side chain R in isohumulone

Splitting off this side chain yields isovaleric acid as the oxidation product. Cleavage of the side chain R from isocohumulone yields isobutyric acid. The structure of the side chain R for isocohumulone is as follows:

$$R = -\overset{\overset{\displaystyle O}{\|}}{C} - CH \overset{\displaystyle CH_3}{\underset{\displaystyle CH_3}{<}}$$

Side chain R in isocohumulone

Acyl side-chain splitting is also involved in the "sun-struck" effect mentioned in chapter 2. When iso-α-acids are exposed to light, photochemical cleavage takes place, yielding 3-methyl-2-butene-1-thiol. This compound is sulfur based and has a pronounced skunky character. It is interesting that the sun-struck phenomenon competes with oxidation, in the sense that beers with high oxygen levels (or with a large amount of oxidized components) tend to be more resistant to photochemical transformations than beers in a reduced state.

Another class of beer-staling constituents consists of fatty acids. In beer, fatty acids come from two sources, namely, unsaturated fatty acids from wort trub and saturated ones from yeast metabolism. As discussed in chapter 3, the saturated fatty acids can react with alcohols to form esters. The unsaturated fatty acids, on the other hand, are major players in beer staling. They tend to be fairly resistant to oxidation and spill over into the finished beer where they tend to produce "fatty or goaty notes."

Fatty acids undergo auto-oxidation by the nonenzymatic pathway. The fate of oleic acid in the nonenzymatic mechanism is typical. This unsaturated fatty acid has the structure

$$R - CH_2 - (CH_2)_6 \overset{\overset{\displaystyle O}{\|}}{C} - OH$$

Oleic acid structure

The presence of molecular oxygen causes oleic acid to oxidize through loss of hydrogen; the oxidized product is

$$RCH = CH(CH_2)_5 CO_2H$$

Oxidized oleic acid structure

This oxidation process can occur on both the hot and cold side of brewing. Whatever oxidized oleic acid spills over to the finished beer will tend to impart fatty and/or soapy notes. Melanoidins undergo auto-oxidation by the same mechanism, which is denoted by

$$4\,(HO-C-\textcircled{M})\ \xrightarrow{\ O_2\ }\ 4\,(O-C-\textcircled{M}) + 2\,H_2O$$

Auto-oxidation of a melanoidin

However, this reaction definitely requires heat and, hence, is induced by molecular oxygen present in wort production. The melanoidins oxidized as a part of hot-side aeration are held in check during fermentation and most of maturation by the strong reducing effects of yeasts. In packaged beer, they can start further oxidation by the mechanism of oxidation by electron transfer without molecular oxygen. The melanoidin-induced oxidation of fusel alcohols is a common reaction:

$$2\,(O-C-\textcircled{M}) + {}_-RCH_2OH \rightleftharpoons 2\,(HO-C-\textcircled{M}) + {}_-RCHO$$

| oxidized melanoidin | alcohol | reduced melanoidin | aldehyde |

Oxidation without molecular oxygen

In this reaction, oxidized melanoidin and alcohol yield reduced melanoidin and aldehyde.

The staling aldehydes tend to have a broad range of off-flavors, although metallic and/or grain astringent notes are often detected. Conversely, in a beer that already contains reduced melanoidins, they can act as flavor protectors by absorbing molecular oxygen and preventing it from reacting with alcohols or other easily oxidizable beer constituents. This effect is why dark beers, prepared with low hot-side aeration and high-melanoidin malts, can have extraordinary flavor stability.

Finally, malt- and hop-based phenols react much in the same way as unsaturated fatty acids and melanoidins. The auto-oxidation products, which are generated during hot-side aeration, have a characteristic harsh, grainy character. Once formed, phenolic compounds, melanoidins, unsaturated fatty acids, and iso-α-acids can interact in a highly complex electron-exchange system. The following diagram is an example in which a phenol is oxidized as oxidized fatty acids and melanoidins are reduced. The center circle refers to the liquid medium in which these reactions take place.

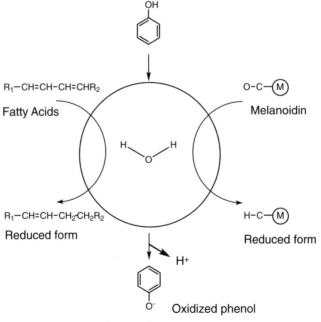

Electron-exchange reaction

Observe that molecular oxygen does not play a role in this system, although the beer flavors that result will invariably be perceived as associated with staling.

Oxidized phenols also display a tendency to polymerize via further oxidation, and this process produces harsh and astringent flavors. Various haze-forming mechanisms accompany this process. One typical mechanism is called Baeyer condensation. Here, wort-derived phenols like catechin react with aldehydes to form a complex.

D-catechin   Acetaldehyde

Both haze and the formation of harsh flavors accompany this reaction. The process is one instance in which haze formation and staling are coupled. Such a mechanism is called an oxidation haze, and it is a good indicator of staling. It, unlike chill hazes, is permanent and will remain even if the beer is warmed.

The role played by hop iso-α-acids in the electron-exchange system is also significant. Most important is the role they play as natural antioxidants (Hashimoto and Kuroiwa, 1981). One such mechanism involves the donation of electrons to melanoidins. This results in the acyl side-chain cleavage discussed above, but it also results in the reduction of oxidized melanoidins. This effect then inhibits the melanoidin-mediated oxidation of alcohols.

Most important, however, is the reduction of unsaturated fatty acids by this same mechanism. This process inhibits the mechanisms (discussed in the next section) that lead to trans-2-nonenal and is the main reason why carefully brewed beers with elevated hop levels are less prone to develop cardboard and/or papery notes compared to other beers. This phenomena is often incorrectly attributed to hop flavors masking the flavors of staling compounds. If anything, the reverse is more common in most beers.

THE SPECIAL CASE OF TRANS-2-NONENAL

The most feared of the staling compounds are a class of long-chain unsaturated aldehydes, of which trans-2-nonenal (T-2-N) is a major example. It has very powerful paper- and/or cardboard-like flavors with a threshold of a mere 0.00010 mg/L (i.e., 0.1 parts per billion). Unlike other staling compounds, there are very few beer drinkers who are taste blind to these compounds, and their presence is universally seen as a major defect. The striking lack of brand loyalty for "import beer" (as defined previously in this chapter) is very likely a reflection of this phenomenon.

Short-chain aldehydes of the form

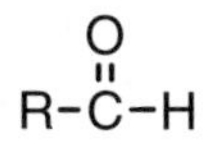

$$\begin{array}{c} O \\ \| \\ R-C-H \end{array}$$

Short-chain aldehyde

are precursors to T-2-N. They, plus whatever T-2-N is present, are called the *T-2-N potential*. These short-chain precursors are somewhat neutral in flavor, but are readily transformed to a longer-chain (and much stronger-flavored) version by aldol condensation (Narziss, 1989). The long-chain aldehydes have the form

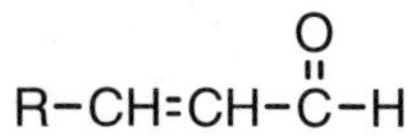

$$\begin{array}{c} O \\ \| \\ R-CH{=}CH-C-H \end{array}$$

Long-chain aldehyde

with T-2-N being the special case, as shown in the following diagram:

$$R = CH_3\text{-}(CH_2)_5$$

Side chain in trans-2-nonenal

It was once thought (see, e.g., Chapon, 1981; Liebermann, 1984; MacFarlene, 1970) that just about all of the oxidative defects, including the presence of T-2-N, were attributable to air in the headspace of beer packages. An early fundamental paper (Hashimoto, 1975) showed that air in the headspace was not the cause of the T-2-N problems and, in particular, that the total T-2-N potential was not affected by the presence of molecular oxygen. If anything, it tends to reverse aldol condensation, although this effect is weak in packaged beer (Narziss, 1989). Later work (Hashimoto, 1986) showed that the T-2-N potential was set by the wort production and changed little from that point forward. These findings were confirmed by Narziss (1990) (see also Huige, 1993; Fix, 1992), who emphasized the elimination of hot-side aeration.

More recent studies (see, e.g., Cantrell and Griggs, 1996; Forster et al., 1998; Lermusuary et al., 1999) have shown that eliminating hot-side aeration is not the entire answer because there were other pathways leading to T-2-N. Confirmation of its presence can be detected during formal tastings of prestigious European beers produced in very modern facilities that have been completely optimized with respect to cold-side aeration and hot-side aeration. On occasion (see Fix and Fix, 1999), the "nonenal problem" with these beers can still be severe.

Dutch researchers (Barker, 1983) showed that lipase and lipoxygenase enzymes present in barley are the key to an enzymatic pathway leading to T-2-N. This pathway along with the hot-side aeration (nonenzymatic pathway) is shown here:

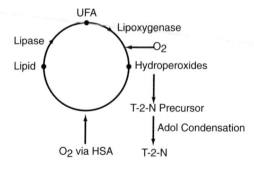

Enzymatic and nonenzymatic pathways

Both pathways occur in wort production and involve unsaturated fatty acids in a fundamental way. It has been shown (Barker, 1983; Nyborg et al., 1999) that in fresh beers, both T-2-N and its precursors are bound up with natural sulfur compounds from yeast metabolism. However, after a lag, which is reduced if thermal or mechanical abuse occurs, the effects of T-2-N start becoming discernible.

Note that the enzymatic pathway does require some oxygen in conjunction with the action of lipoxygenase. However, this requirement is small (typically in the range of parts per billion) and thus virtually impossible to avoid in any type of brew house, even modern high-tech configurations that have been completely optimized with respect to hot-side aeration. As a consequence, the practical control of T-2-N has focused on wort lipid levels and the malts used. First of all, achieving a reasonable wort clarity in lautering is important. It is interesting that some of the biggest nonenal problems have been in breweries who have stepped up to a 12-brew-a-day schedule, and with this increase, their wort has undergone a reduction in clarity (Fix and Fix, 1999).

That malt and malting are involved in the formation of T-2-N was first suggested by Doperer et al. (1991). It is true that lipase and lipoxygenase are thermally unstable and thus 96% of these enzymes are lost on kilning of pale malts (Cantrell and Griggs, 1996) in a process called blanching. Even more is lost for Munich and Vienna types. Unfortunately, because of the very low flavor threshold of T-2-N, not much is needed before the deleterious flavor appears. As a consequence, careful malt selection (and reasonable wort clarity) may be the best way to minimize the effects of T-2-N, assuming of course that hot-side aeration is not a factor.

Beer Stabilization

5

Some degree of clarity is usually considered to be a desirable feature in most beers. This can range from the brilliant clarity of a well-made Pilsener to the translucent haze found in cast-conditioned ales and German hefeweizens. Even with dark and essentially opaque beers, a well-defined "blackness," in contrast with a murky and/or muddy appearance, is normally an essential feature of the beer's overall quality. The goal of this chapter is to lay out the options available to brewers for achieving their desired end result.

MECHANISMS LEADING TO BEER HAZE

The most common, and in some sense, the most important type of beer haze—since it is relevant to many beer types—is chill or temporary haze. As the name suggests, this haze appears when the beer is suitably chilled; the haze disappears upon warming. The temperatures at which the haze appears and disappears depend on the physical stability of the beer. The more stable the beer, the closer to 0 °C before chill haze occurs. The haze involves complexes of high-molecular-weight proteins and polyphenols (tannins). These compounds form weak, temperature-sensitive hydrogen bonds that are broken as the beer's temperature increases, allowing the resulting compounds to form a complex with water molecules and go into solution.

The other forms of haze are permanent; they are characterized by strong bonds of covalent type in which constituent atoms share the available electrons to achieve a more stable energy state. All forms of permanent haze invariably arise from technical errors in brewing. The most important kinds of permanent haze are starch haze, biological haze, and oxidation haze, but there are other types of haze mechanisms as well.

Large-molecular-weight carbohydrates—including gums like ß-glucans—will fall out of solution and form an intractable haze. This effect primarily comes from problems in mashing and sparging. Less starch breakdown than desirable during mashing and the resulting extraction of high-molecular-weight carbohydrates in the sparge are the chief sources of starch haze. Poor-quality malt, especially malt with high ß-glucans levels, may also be indicated. Because all of these problems are avoidable, starch haze will never occur in beers made with quality malt and a rational mashing and sparging program.

Many microbes hostile to beer will form rope (i.e., suspended thin strips of particles in beer) and haze. As a general rule, this problem will occur only in very serious contaminations, typically many thousands of cells per milliliter. Off-flavors due to contamination will be detected at much lower levels. As a consequence, clarity does not always indicate that the beer is free of the deleterious effects of wild yeasts and acid-forming bacteria. Another type of microbial haze is formed by yeasts that have not fully flocculated by the end of fermentation. Minor yeast haze can readily be corrected by filtration and fining, which is discussed in the next section. Serious yeast haze can be a signal of undesirable yeast mutation.

As beer ages, free oxygen or oxidized compounds having O_2 bound up with other constituents can cause intractable proteins and/or polyphenol hazes. These hazes will arise in all beers, but those prepared from low-oxygen systems tend to last many months before an oxidation haze is evident.

Abuse in trade is a term that applies to undesirable levels of heat, light, and mechanical abuse. All of these mistakes in handling will create haze in otherwise well-made beers. Beers made with water containing higher than acceptable metal levels (see Fix and Fix, 1997) are particularly vulnerable. The same is true of bottled beer having high headspace air levels.

In this chapter, haze is specified in terms of FTUs (formazin turbidity units). Intuitive descriptions of the important ranges are given in Table 5.1 along with the analogous EBC (European Brewery Convention) units. Easy-to-follow protocols for measuring FTUs are in *Methods of Analysis* (ASBC, 1987). The procedure formulated by the ASBC consists of visual comparison of beer with a working-standard colored solution. The latter is prepared from formin (hexamethylenetetramine) and hydrazine sulfate. The determination is done by diluting colorless distilled water with known quantities of the working standard until a visual match is made with the sample of beer. FTU determination can also be automated by use of suitably calibrated haze meters. The latter are rather expensive and do not offer a qualitatively significant increase in accuracy over visual methods.

Table 5.1	**Comparison of Beer-Haze Units**	
FTUs	**EBC units**	**Visual effect**
0–100	0–1.5	brilliant clarity
100–200	1.5–3	clear with slight dullness
200–300	3–4.5	see-through haze
>400	>6	murky

A number of procedures have been proposed for the determination of the physical stability of newly bottled beer (Berg, 1991). A widely known procedure is the so-called "warm day analysis." One warm day consists of incubation of the beer sample at 60 °C for 48 hours followed by incubation at 0 °C for 24 hours. This procedure is used to evaluate the effectiveness of several beer-stabilization methods in this chapter. The procedure consists of treating a sample of beer for 10 warm-day cycles, after which a subsample of the beer is then treated for five more cycles. At the end of each cycle, the FTU of the beer is determined by the ASBC visual method (see Table 5.1). Ideally, one would want the beer to be below 200 FTUs at the end of 10 warm days and, at the other extreme, under 300 FTUs after 15 warm days.

This type of analysis also has many benefits other than the evaluation of stabilization techniques. For example, it is a measure of malt quality as well as wort-production techniques.

Another scheme that has been found to be very useful in the prediction of beer-haze stability under standard storage conditions is to use a lower stressing temperature, namely, 37 °C. A good rule introduced by T. O'Rourke (1996) is that one week at 37 °C is equivalent to one month at 18 °C. Haze measurement using the ASBC method is done at 37 °C and 0 °C. These two procedures are used in this chapter to evaluate the various stabilization procedures.

FINING AGENTS

There are three major types of fining agents: chill-haze–proofing agents that remove high-molecular-weight polyphenols, chill-haze–proofing agents that remove high-molecular-weight proteins, and fining agents that reduce the yeast biomass present but that generally do not affect colloidal chill-haze stability.

Balance is the key for any effective fining protocol. When overdone, all of the fining agents can create more problems than they solve. In addition, there is the point raised in the preceding section that neither fining nor filtration is a practical way to deal with hazes arising from technical problems in brewing.

POLYPHENOL AGENTS

There are two reasons why special attention has been given to fining agents that work on the phenol side of the polyphenol-protein haze complex. First, these agents do not affect beer foam, something that can be a problem for those working on proteins. Second, phenols play a special role in chill haze. In addition to high-molecular-weight phenols (tannoids), even simple phenols get into the fray. As beer ages, the simple phenols polymerize, and the higher-molecular-weight compounds that result become available for haze formation. This effect is clearly shown in Figure 5.1 for beer stressed at 37 °C. As the beer ages, the concentration of simple phenols decreases, and this process occurs simultaneously with an increase in tannoids and haze. This aging process shows why all beers will ultimately be susceptible to chill haze, but by removing the high-molecular-weight components at the start of storage, this effect can be delayed to give the beer more shelf life.

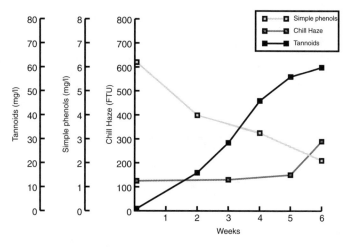

Figure 5.1. Storage of untreated beer at 37 °C.

The most commonly used fining agent that works on polyphenols is PVPP (polyvinylpolypyrrolidone). The amount added is of great importance. As illustrated in Figure 5.2, a dosage of 10 g/hL has virtually no effect, i.e., the amount of chill haze produced over time with the 10 g/hL dosage is the same as that produced with no treatment (cf. Fig. 5.1). The best results came from a dosage of 50 g/hL, which is widely regarded as optimal (McMurrough et al., 1997). An interesting effect (first reported in O'Rourke, 1996) is that with a dosage of 50 g/hL, there is virtually no change in the tannoid concentration as the beer ages, in contrast to the increases seen in untreated beer at 10 g/hL. The measurements illustrated in Figure 5.3 show that PVPP has the ability to selectively hydrogen bond just about all haze-relevant polyphenols. Dosages in excess of 50 g/hL are not recommended since they do not improve haze stability over what is achieved at a concentration of 50 g/hL. In addition, excess PVPP levels can reduce color and hop flavor.

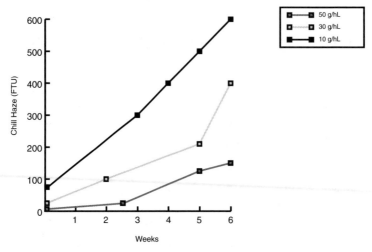

Figure 5.2. Storage at 37 °C after PVPP treatment.

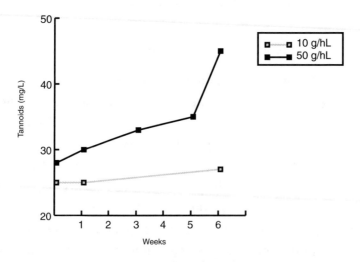

Figure 5.3. Evolution of tannoids at 37 °C.

PROTEIN AGENTS

Silica gels and sols are also widely used because they delay chill haze and, at the same time, do not affect beer foam if properly applied. They work via an absorption mechanism that is remarkably selective to haze-forming proteins. Their impact is illustrated in Figure 5.4, which was adapted from Siebert and Lynn (1997).

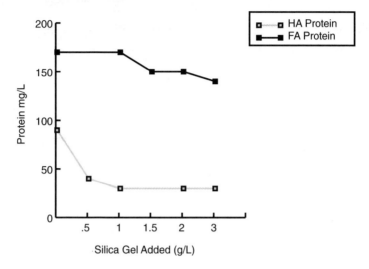

Figure 5.4. Effectiveness of silica gel to remove foam-active and haze-active proteins.

Brewers who need extended shelf life for their products (e.g., more than 6 months) will generally use both silica gel and PVPP since these two agents work independently of each other. Scheer (1990) discussed various options for such combinations.

One of the oldest protein-active fining compounds is Irish moss. It has been used throughout the twentieth century in both the United Kingdom (Hind, 1950) and the United States (Nugy, 1937), but it has never found favor on the continent. It consists of refined seaweed, and its major components are carrageenan and 3,ß-anhydrogalactose. It is available in commercially prepared products, which tend to vary. All, however, attract protein by their electrical charge. Proteins can have either a positive or negative charge, and their charge depends fundamentally on the pH of the medium. The haze-active proteins are electrically neutral at a pH of about 6.0 (the so-called isoelectric point) and carry a positive charge for a pH below this. The pH of boiling wort is well below 6.0; thus, in this environment, the negatively charged Irish moss will electrically bind with the proteins and drop out of solution. Unfortunately, part of the amino acid pool as well as some foam-active proteins may carry a slight positive charge, and at too high a dosage, Irish moss could remove some of these as well. This effect is illustrated in Figure 5.5 for refined Irish moss flakes. The latter were hydrated in water and added 15 minutes before the end of the boil.

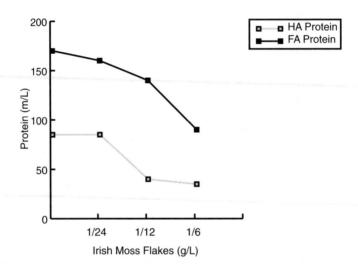

Figure 5.5. Effect of Irish moss flakes.

YEAST-ACTIVE AGENTS

Yeast-active compounds are used to reduce yeast biomass and typically will not have any appreciable chill-haze–proofing capability. Originally, these agents were used in cast-conditioned ales, although increasingly, lager brewers are using them before filtration to reduce the biomass loading on the filter (Fix and Fix, 1997). Isinglass is the most popular yeast-active fining compound. It comes from the swim bladders of tropical fish, and its chief ingredient is collagen. Fix and Fix (1997) presented instructions for its preparation; its effect is shown in Figure 5.6.

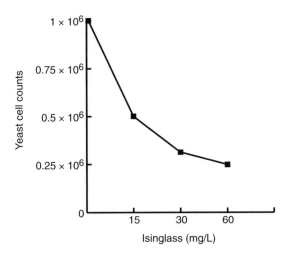

Figure 5.6. Yeast biomass decrease with isinglass use.

ICE STABILIZATION

There has been a growing interest in natural, nonadditive methods for beer stabilization. First, brewers have traditionally seen beer as a natural organic libation, and the fewer things that are added to it—above and beyond grains, hops, yeasts, and water—the better. Changing labeling laws is another factor. In the past, a long list of compounds were considered as "processing agents." All of the fining agents discussed in this chapter are examples. They can be added during brewing, and at the same time, the beer label can declare "no additives." This approach is different from that taken for other food products and, hence, is subject to change in the future. Rather than using additives with advertising campaigns declaring "made with the finest seaweed" (Irish moss) or "made with the finest plastic" (PVPP), brewers have been seeking other solutions that do not involve such compounds!

Cross-flow cartridge filtration falls into this category, since additive supplements like diatomaceous earth are not required. However, filtration of any type by itself cannot satisfactorily produce chill-haze–proof beer. As will be seen in the next section, the material trapped on even submicrometer filters consists of biomass and gums like αglucans. Only minor amounts of haze-active protein are trapped.

Ice stabilization is an important nonadditive supplement to filtration, which is capable of effectively chill-haze–proofing beer. The traditional "ice beer" process consists of partly freezing the beer during maturation. In North America, this process is classified by law as *distillation* since water is removed and therefore the concentration of alcohol is increased. Such distillation processes have been used for centuries to make high-alcohol bock beers. In addition to their strength, anecdotal evidence suggests that such beers have two key attributes. First, they tend to have remarkable flavor stability, far greater than standard bock beers. Second, they tend to be very smooth because all "rough edges" in their precursors have been removed by the freezing process (Kunze, 1996).

Recently, the Labatt group came up with an unique and highly original variation of ice stabilization that does not involve distillation. Key to their research is the discovery that the benefit from ice stabilization is a highly nonlinear function of temperature. In particular, there is a narrow temperature range below the freezing point of beer in which the maximum benefits are achieved, as illustrated in Figure 5.7.

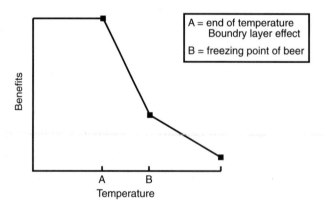

Figure 5.7. Temperature effect in ice stabilization. Point A represents the freezing point of beer; point B represents the lower temperature of the temperature range over which maximum stabilization benefits occur. .

The temperature at which beer actually freezes depends fundamentally on its alcohol content and to a lesser degree on the residual extract. Table 5.2 gives some typical values. The lower end of the temperature range, i.e., the point B in Figure 5.7, is certainly warmer than –5 °C and generally about –4 °C.

Table 5.2	**Approximate Freezing Temperatures**	
Alcohol (% w/w)	**RE (°P)**	**Freezing temperature (°C)**
4	3.5	–2.3
4.5	4.0	–2.5
5	4.5	–2.7
5.5	5.0	–2.9

The industrial ice-stabilization process (U.S. Patent No. 5,304,384, Labatt Brewing Company, April 19, 1994) is shown schematically in Figure 5.8. The brewed green beer is rapidly cooled in a scraped-surface heat exchanger to a temperature between –2 °C and –4 °C. Very small ice crystals are formed that are less than 5% of the total volume. This restriction is an important point, since ice stabilization is not a distillation or concentration process, i.e., there is only a minimal increase in the alcohol level. The cooling beer and ice crystals are then moved into a container called the *ice zone*. This slurry is mixed with more with ice crystals and green beer and is maintained in a constant state of agitation to render it homogeneous. This technology permits the rapid treatment of beer. For example, the time in the scraped-surface heat exchanger is less than 60 seconds, and the time that it spends in the ice zone is from 5 to 15 minutes. Then the beer is extracted from the ice and beer mixture and is processed with standard procedures.

Figure 5.8. Industrial ice stabilization.

An alternative to this process is a single-tank version, in which the partial freezing is done in maturation tanks. The only disadvantage of this process is that it is not particularly amenable to mass production and longer times are needed. A general rule is that 72 hours at –3 °C will keep ice formation below 5%, although process conditions and beer composition are confounding factors. The following two points are important to follow in the tank version: (1) Yeast-cell count should be kept below 100,000 cells per milliliter since the icing effect can impart a "yeast bite." (2) The beer should be transferred off the ice before it thaws and nullifies some of the benefits of the procedure.

The documented benefits to ice stabilization are twofold. First, it acts on both haze-active phenols and proteins. Thus, it is a substitute for both PVPP and silica gels. Test brews have shown there is about a 5% decrease in simple phenols, tannins, and haze-forming proteins over what is normally achieved in fining procedures using PVPP or silica gel. Yet remarkably, data presented in the next section shows that there is about a threefold increase in chill-haze proofing. This result is indirect evidence that slight ice formation is more selective in the removal of haze-forming compounds than exogenous absorbents.

Another feature of ice stabilization, which has long been suspected from anecdotal evidence, is that it is also selective to polyphenols in the higher oxidation states. Indicator time tests (I.T.T., a method for measuring a beer's degree of oxidation, as explained in Chapon, 1981, with improvements in Rabin and Forget, 1998) show that there is about a 30%–40% reduction in the time required to decolorize 80% of a standard indicator (2:6 dichlorophenol/indophenol). This decolorization test is not directly quantitatively equivalent to a given reduction in the beer's redox state; nevertheless, in practical terms, there is a strong correlation between the two (Chapon, 1981). PVPP will also reduce a beer's oxidation state, but not nearly to the extent of ice stabilization.

FILTRATION

Historical references show that brewers have been using some type of filtration for centuries (Arnold, 1911). If properly used, it can serve as an effective nonadditive tool in beer clarification. However, it is important to note that any well-made beer using top-quality malt and yeasts should clarify itself during postfermentation maturation. In this regard, filtration, and indeed fining, should not be seen as a means of correcting haze arising from technical errors in brewing. As noted in the first section of this chapter, the best practical step in these cases is to correct the original problem that created the haze. Filtration is therefore best seen as a final polishing of beer, used in conjunction with fining to render it brilliantly clear and stable with respect to temperature changes. Filtration can be employed on one of the following levels:

- supermicrometer filtration
- 1 μm filtration
- submicrometer filtration

There are actually two ranges used in supermicrometer filtration, namely, trap filtration at the 10–20 μm range and the more conventional 3–5 μm filtration. Trap filtration is used in "unfiltered beer" to trap most of the sediment in maturation tanks. It is also used in ice stabilization to separate the crystals formed at subfreezing temperatures. Filtration in the 10–20 μm range will allow virtually all microbes to penetrate, but will not alter the hop, protein, and carbohydrate concentrations.

Filtration in the 3–5 μm range will trap the sediment formed during filtration. Its most notable effect is the removal of dead and dormant yeast cells, something that gives the filtered beer a cleaner flavor. Cell counts reported in Fix and Fix (1997) always fell below 100 microbes per milliliter of the filtered beer. As for filtration through coarser filters, filtration at 3 μm has virtually no effect on beer composition (see Table 5.3). Therefore, this is the ideal filter for unpasteurized beer that is not subject to market abuse.

Brewers producing locally distributed unpasteurized beer for which a shelf life of several months is important will typically filter at the 1 μm level. This will eliminate all yeasts in the packaged product. However, large numbers of beer-spoiling microbes, if present, will penetrate such filters. As a consequence, nationally marketed products are usually filtered at 1 μm and pasteurized. Compared to unfiltered beer, there are some modifications of the beer's composition during 1 μm filtration, as indicated in Table 5.3.

Sterile filtration is an alternative to pasteurization; however, this procedure requires filtration at a level of 0.5 μm or finer. This level of filtration still will not trap all microbes. A filter as fine as 0.22 μm is needed for that (Ryder et al., 1988). Nevertheless, no microbe hostile to beer has been identified after a proper filtration at 0.5 μm or finer. The composition of beer, on the other hand, is changed in nontrivial ways with filtration through finer filters, and this fact must be taken into account for the more flavorful beer styles.

The bottom line for each brewer is to find the combination of filtration (if any) and/or fining (if any) that best serves the brewer's goals for the particular beer style under consideration. Essentially, there are the five methods shown in Figure 5.9. To compare the five options, the following was used:
- Filtration: 3 μm cartridge filtration
- Fining: 30 g/hL PVPP and 20 g/hL silica gel
- Ice stabilization: 2% volume reduction

The data in Figure 5.9 show that the no-filtration–no-fining procedure (curve a) typically will render a moderately hazy beer that gets murkier with time. Fining without filtration (curve b) will delay the increase in haze. Filtration without fining (curve c), on the other hand, will typically lead to a brilliantly clear beer, but one that is not chill-haze proofed. Predictably, fining with filtration (curve d) will delay haze formation.

Table 5.3	**Filtration Data**			
	Values for unfiltered beer	**Values after filtration through membrane with rating of**		
		0.45 μm	**1.0 μm**	**3.0 μm**
Real extract (°P)	4.7	4.4	4.8	4.9
Protein (%)	0.72	0.65	0.66	0.70
Color (°Lovibond)	4.6	3.7	3.8	4.6
Bitterness (BU)	15.1	14.4	13.8	14.1
NOTE: From Fix and Fix (1997) and Siro and Lovgrem (1980).				

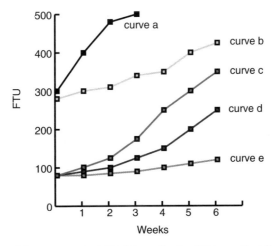

Figure 5.9. Five stabilization strategies. (a) No fining or filtration. (b) Fining without filtration. (c) Filtration without fining. (d) Filtration with fining. (e) Filtration with ice stabilization.

By using the rule that one week's storage at 37 °C is equal to one month's storage at 18 °C, the data in Figure 5.9 indicate that fining with filtration will typically render a beer stable over a 4-month interval. Finally, ice stabilization (curve e), as anecdotal evidence has suggested, produces beers with truly remarkable haze stability.

Gases

6

IDEAL GASES

The three most important gases in brewing are CO_2 (carbon dioxide), N_2 (nitrogen), and O_2 (oxygen). Air is also important; however, it can be treated as an $O_2 + N_2$ mixture. Atmospheric air is approximately 21% oxygen and 78% nitrogen plus trace gases (most notably argon).

The kinetic molecular theory (Vincenti and Kruger, 1982) provides the most useful model of gases as far as brewing is concerned. It is based on the following assumptions:

- The gas consists of a large number of very small particles called *molecules*. Each molecule moves at a constant speed in a straight line. Different molecules can have different velocities.

- The gas molecules are so small that they can be assumed to have no volume. The volume of the gas is, thus, the volume of the container it is in.

- The gas molecules act like billiard balls in the sense that they collide and thereby change directions. Collisions between two molecules are elastic in the sense there is an exchange of kinetic energy (one moves faster than before and the other slower), but the total kinetic energy is conserved.

- The absolute temperature (which is measured from absolute zero [−273.15 °C] in kelvins, K) is proportional to the average kinetic energy of the ensemble of gas molecules. This statement means that as the temperature increases, the gas particles move faster.

- The gas molecules also have collisions with the container walls, and the result of these collisions constitutes the pressure in the system. The greater the number of collisions per unit time, the greater the pressure. The average pressure of the atmosphere at sea level is referred to as *one atmosphere* (1 atm). It is equal to 14.7 psi (pounds per square inch).

In addition to volume, pressure, and temperature, the concept of a *mole* is also important to the study of gases, whether in beer, the atmosphere, or wherever a gas occurs. A mole is a set number of molecules of any substance. The substance could be an elemental substance such as argon (Ar) or gold (Au), or it could be a compound such as oxygen (O_2), nitrogen (N_2), water (H_2O), carbon dioxide (CO_2), or ethanol (C_2H_5OH). For any elemental substance, such as gold, 1 mol (*mol* is the SI symbol for *mole*) contains 6.02×10^{23} atoms. For any compound, 1 mol contains 6.02×10^{23} molecules. This large number is called *Avogadro's number.*

The mass of 1 mol of a compound is equal to its molecular weight in grams (g), e.g., 1 mol of CO_2 has a mass of 44 g; 1 mol of N_2 has a mass of 28 g; and 1 mole of O_2 has a mass of 32 g. The density of a compound is defined as the ratio of mass to volume, e.g., the density of water is 1 g/mL and the density of ethanol is 0.7893 g/mL. The density is distinguished from the specific gravity of a compound; the specific gravity is the ratio of the density of the compound to the density of water at a standard temperature. Thus the specific gravity of ethanol is 0.7893/1 = 0.7893 (the units cancel out).

Thus, there is a relationship between the number (n) of moles—i.e., the amount—of a substance, the mass (m) of the substance in grams, and the molecular weight (MW) of the substance:

$$n = m/\text{MW} .$$

(6.1)

For example, how many moles are represented by 50 g of CO_2? As stated, the molecular weight of CO_2 is 44 g/mol, so

$$n = (50 \text{ g of } CO_2)/(44 \text{ g/mol}) = 1.14 \text{ mol of } CO_2 \, .$$

Conversely, 1.14 mol of N_2 has a mass of

$$m = (28 \text{ g/mol}) + (1.14 \text{ mol of } N_2) = 31.92 \text{ g of } N_2 \, .$$

Finally, 44 g of CO_2, 28 g of N_2, and 32 g of O_2 all have the same number of molecules (by definition, because each amount is 1 mol)—namely, the Avogadro's number.

At the pressures and temperatures typical of brewing, CO_2, N_2, and O_2 behave essentially as ideal gases. For these, volume V (in liters, L), pressure P (in atmospheres, atm), absolute temperature T (in kelvins, K), and number of moles n (in moles, mol) satisfy the relationship

$$(PV)/(nT) = R \, .$$

R has a constant value and is called the *universal gas constant*. The value of R has been shown to be 0.0821 (L × atm)/(K × mol). Rewriting this equation gives the *ideal gas law*:

$$PV = nRT \, . \tag{6.2}$$

For example, suppose that a 25 L tank holds 50 g of CO_2, and suppose that the temperature is 10 °C. Rearranging equation 6.2 predicts that the pressure in the tank can be calculated as follows:

$$P = (nRT)/V \, .$$

Since the mass of CO_2 is 50 g, the number of moles (from equation 6.1) is

$$n = (50 \text{ g})/(44 \text{ g/mol}) = 1.14 \text{ mol} \, .$$

In addition, because the absolute temperature (in K) is the temperature in Celsius (°C) plus 273 °C (actually 273.15 °C), the absolute temperature is

$$T = 10 \text{ °C} + 273 \text{ °C} = 283 \text{ K} \, .$$

The temperature units °C and K are the same size, i.e., they represent the same change in temperature. The two scales that use them—the Celsius and kelvin (or absolute) temperature scales—just start at different points, 0 °C and 0 K, respectively, and 0 K = –273.15 °C. Thus,

$$P = [(1.14 \text{ mol}) \times (0.0821 \text{ (L (atm)/K} \times \text{mol))} \times (283 \text{ K})]/(25 \text{ L}) = 1.06 \text{ atm} .$$

There are three special cases of the ideal gas law that are expressions of how a fixed mass of gas (i.e., a fixed number of moles n) changes with changes in pressure, volume, and temperature. For this fixed-mass situation, combining two statements of the ideal gas law (i.e., equation 6.2) for a sample of gas at two different conditions yields

$$(P_1V_1)/T_1 = (P_2V_2)/T_2 . \tag{6.3}$$

Note that n and R drop out of the calculation. In equation 6.3, P_1, V_1, and T_1 are the pressure, volume, and absolute temperature (in K) of any sample of gas. When changes—e.g., heating or cooling—are made to the sample, new values P_2, V_2, and T_2 result, and these are related to the old values by equation 6.3. The three special cases are shown in Table 6.1.

In a continuation of the preceding example, suppose that the temperature of the CO_2-filled tank is increased from 10 °C to 20 °C. Because a tank is involved, instead of a flexible container such as a balloon, the volume is constant. Table 6.1 indicates that a constant-volume situation is covered by Gay-Lussac's law. Applying the relationship

Table 6.1	Three Gas Laws for a Fixed Mass of Gas	
Name	Equation	Constant
Boyle's law	$P_1V_1 = P_2V_2$	$T_1 = T_2$
Charles' law	$V_1/T_1 = V_2/T_2$	$P_1 = P_2$
Gay-Lussac's law	$P_1/T_1 = P_2/T_2$	$V_1 = V_2$

shown predicts an increase in pressure, as follows:

$$T_1 = 10\ °C + 273\ °C = 283\ K\ ,$$

as already determined, and

$$T_2 = 20\ °C + 273\ °C = 293\ K\ .$$

Also, as already determined,

$$P_1 = 1.06\ atm\ .$$

From Gay-Lussac's law, it follows that

$$P_2 = (T_2 P_1)/T_1 = [(293\ K) \times (1.06\ atm)]/(283\ K) = 1.10\ atm\ .$$

APPLICATIONS OF THE LAW OF PARTIAL PRESSURES TO BREWING

In addition to water vapor, mainly produced by vaporization during the boil and not the focus of this chapter, gases in beer are actually present as a mixture of CO_2, N_2, and O_2. Thus it might seem that the ideal gas law would not apply. Fortunately, it does! John Dalton was an English chemist and physicist who lived from 1766 to 1844, and the major feature of his work showed that in a mixture of gases, each molecule acts independently of all others. This statement assumes that the gases do not interact with each other as they might when they are highly compressed. As far as beer is concerned, neither compression nor vaporization is sufficient to void Dalton's statement, and so this assumed independence of individual gas molecules is a very accurate picture of what actually occurs.

What this concept means is that in a mixture of gases, each component has its own separate $PV = nRT$ relationship and is independent of the action of the other components. The pressure P of an independent component is called the partial pressure of that component. It is the pressure that the component would exert if it were the only gas in the system. The total pressure P is then the sum of the individual partial pressures:

$$P = P_{CO_2} + P_{N_2} + P_{O_2}, \qquad \text{(6.4)}$$

where the individual pressures on the right side of equation 6.4 are the partial pressures of CO_2, N_2, and O_2, respectively. This concept is called *Dalton's law of partial pressures*.

In the next three sections, three applications of this basic law are considered. The first deals with the carbonation of beer. The second involves the oxidation of beer, and the last explores the implications of Dalton's findings for dispensing beer. The dispensing considerations are particularly important to brewers. Indeed, there is nothing more frustrating than to produce a high-quality beer, one for which all the technical issues discussed so far in this book have been correctly resolved, only to have the beer compromised by the way it is dispensed.

CARBONATION OF BEER

The unit generally used for measuring the amount of CO_2 dissolved in beer is called *volumes* (abbreviated Vol.) in the brewing literature. It is defined as follows:

$$1 \text{ Vol.} = 1 \text{ L of } CO_2 \text{ per liter of beer.}$$

Table 6.2 gives some representative values of the CO_2 content of various types of beer.

Table 6.2	CO₂ Volumes Data
Volumes (Vol., L/L)	Beer type
1.0–1.75	Real ale
2.2–2.4	Draft lager
2.4–2.6	U.K. and continental bottle beers
2.6–2.8	U.S. lager beer
>3.0	Bavarian wheat beers

Since the density of CO_2 at standard temperature and pressure is 1.96 g/L and its molecular weight is 44 g/mol, the number of moles n per liter is determined as follows:

$$n/V = \text{Vol.} \times (\text{density/MW}).$$

$$= \frac{\text{Vol. (in L of } CO_2)}{(1 \text{ L of beer})} \times \frac{(1.96 \text{ g of } CO_2)/(\text{L of } CO_2)}{(44 \text{ g of } CO_2)/(\text{mol of } CO_2)} \quad (6.5)$$

$$= (\text{Vol.}/22.45) \text{ mol of } CO_2 \text{ per } 1 \text{ L of beer}.$$

The unit for this kind of expression—the number of moles of a gas per 1 L of liquid—is called the *molarity* of the gas in the solution. For example, a beer at 2.5 volumes has a molarity of

$$n/V = 2.5\text{Vol.}/22.45 = 0.11 \text{ mol/L}.$$

Molarity is often expressed by a capital M, e.g., 0.11 M.

The gas laws hold for CO_2 in the headspace of a tank or bottle. In particular, if equation 6.5 is mathematically rearranged and P/RT is substituted for n/V (the ideal gas law),

$$\text{Vol.} = 22.45(n/V) = 22.45(P/RT) \quad (6.6)$$

can be used to compute the molarity and hence the amount of CO_2 in the headspace given the partial pressure P_{CO_2} and the temperature T (in K).

For the CO_2 actually dissolved in the beer, the situation is considerably more complicated because of gas and liquid interactions. Fortunately, a relationship called Henry's law applies to this situation. It states that the concentration of a slightly soluble gas (e.g., CO_2) is directly proportional to the partial pressure of the gas. Moreover, when a mixture of two gases (e.g., CO_2 and O_2) is in contact with a solvent (e.g., beer), the amount of each gas that is dissolved is the same as if it were present alone at a pressure equal to its own partial pressure. By using Henry's law, values for n/V, and hence volumes

(Vol.), for various partial pressures P and temperatures T have been worked out (ASBC, 1987). This often-quoted data set is given in part in Table 6.3. For example, if the partial pressure is 14.7 psi = 1 atm and the temperature is 0 °C = 273 K, then by interpolating it can be seen that Vol. = 3.21.

To give a specific case, suppose that a brewer wants to carbonate 100 L of beer in a vessel of 125 L capacity. The 20% headspace is common for carbonation tanks. In addition, suppose that 2.5 volumes of CO_2 is required in the finished product. Since approximately 0.2 volumes is lost at the filler, the brewer's target should be 2.7 volumes. This result means that, under equilibrium conditions,

$$(2.7 \text{ Vol.}) \times (1.96 \text{ g/L}) = 5.29 \text{ g of } CO_2 \text{ per liter of beer}$$

is required, or a total of

$$(5.29 \text{ g/L}) \times 100 \text{ L} = 529 \text{ g} = 0.529 \text{ kg} .$$

Suppose the temperature of the tank is held at 2 °C = 275 K, which is the same as 35 °F. The data in Table 6.3 indicate that to contain 2.7 volumes of CO_2 at 2 °C, an equilibrium partial pressure of CO_2 of 12 psi is needed. It must be emphasized that time is required for the system to reach equilibrium, and none of the analysis in this example is applicable until the steady state is reached. For direct injection, this time can be significantly reduced by carbonating stones, i.e., diffusers that release CO_2 in a spray of tiny bubbles. If the beer is carbonated by addition of new wort (i.e., kraeusening), then equilibrium is usually established shortly after the second fermentation is complete in gas-tight containers.

To compute the total amount of CO_2 added, the ideal gas law can be used to determine the amount of CO_2 in the tank headspace. Henry's law shows that the partial pressure of the CO_2 in the headspace affects the amount of CO_2 in the beer. Thus

$$P_{CO_2} = (12 \text{ psi})/[(14.7 \text{ psi})/\text{atm}] = 0.82 \text{ atm.}$$

Table 6.3 **Values of CO$_2$ volumes as a function of pressure and temperature**

	4 psi	8 psi	12 psi	14 psi	16 psi	20 psi	33 psi	34 psi	35 psi	36 psi
0 °C	2.05	2.48	2.90	3.11	3.30	–	–	–	–	–
2 °C	1.87	2.29	2.67	2.86	3.05	–	–	–	–	–
6 °C	1.63	1.99	2.34	2.52	2.69	3.04	–	–	–	–
10 °C	1.41	1.74	2.06	2.21	2.38	2.70	–	–	–	–
25 °C	–	–	–	–	1.58	1.72	2.37	2.42	2.47	2.52

NOTE: Where no data (–) are listed, the conditions are not relevant to brewing.

Equation 6.6 predicts that the volumes of CO_2 in the headspace is

$$Vol. = (22.45 \text{ mol of CO2})/(\text{L of beer}) \times (0.82 \text{ atm})/[(0.0821$$
$$L(atm)/(K(mol)] \times (275 \text{ K}) = 0.81 \text{ volumes,}$$

or

$$(0.81 \text{ Vol.}) \ (1.96 \text{ g/L}) = 1.6 \text{ g/L}$$

Since the tank headspace is 20 L, there are

$$(1.6 \text{ g/L}) \times 20 \text{ L} = 32 \text{ g of CO}_2$$

in the headspace. Thus a total of $529 + 32 = 561$ g of CO_2 should be added to the tank. These calculations are summarized as follows:

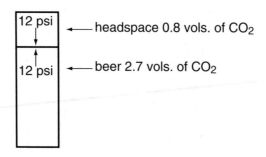

12 psi — headspace 0.8 vols. of CO_2

12 psi — beer 2.7 vols. of CO_2

CO_2 distribution in beer and headspace

OXIDATION OF BEER

As a second application of Henry's law, consider the role played by air in packaged beer headspace. Suppose there is a bottle containing 350 mL = 0.35 L of beer with a headspace of 20 mL = 0.02 L. Suppose, that after filling, it was determined that the beer's headspace air level was 0.5 mL = 0.0005 L. Dalton's law applies to this situation independent of how much CO_2 is involved. In particular, if the version of the gas law due to Boyle (see Table 6.1) is used, the ratio of the partial pressure of air P_{air} to a reference pressure P_{ref} is proportional to the ratio of the volumes of air V_{air} in the headspace to the volume of the headspace $V_{headspace}$, i.e.,

$$P_{air} = (V_{air}/V_{headspace}) \times P_{ref}.$$

Following the procedures found in *Methods of Analysis* (American Society of Brewing Chemists, 1987), $P_{ref} = 1$ atm. Therefore,

$$P_{air} = (1 \text{ atm}) ([(0.5 \text{ L})/(20 \text{ L})] = 0.25 \text{ atm} .$$

Partial pressure of air = 0.023 atm

Partial pressure of air = 0

Partial pressure of air in headspace and in beer

Initially, the gases in the headspace will stratify according to weight; i.e., the 0.5 mL volume of air will sit above a CO_2 layer. Henry's law, however, asserts that, at equilibrium, the gases will act as if each were the only gas present. Since the beer was assumed to be free of air at fill, the system is not at equilibrium because there is a concentration differential between the headspace and beer. This forces an air ingress into the beer. If the air were inert in beer, the ingress would continue until the partial pressures balanced. However, beer constituents (e.g., ethanol) will react with the free oxygen in the ways discussed in chapter 4, and the reactions in turn will reduce the partial pressure of the air in the beer, causing more air to diffuse into the beer until ultimately all headspace air is consumed.

It should be noted that the headspace partial pressure, $P_{headspace}$, increases proportionately with the storage temperature (i.e., absolute temperature, in K). This relationship in part explains why the staling rate increases with temperature. As the pressure increases, air is forced into the beer at a greater rate. Since the rate of redox reactions increases exponentially with temperature, the oxygen-bearing components that are absorbed are consumed at greater rates as well.

When one is doing very accurate CO_2 measurements, it is necessary to take into account the amount of headspace air just after fill. The above considerations apply here as well. As an example, suppose that a standard air and CO_2 tester is used (the tester from the Zahm and Nagle Catalog, Buffalo, New York, for example, which comes with an expanded version of Table 6.3). This is a piercing device that measures the total pressure (i.e., CO_2 plus air) in the headspace, the headspace air volume, and the temperature. Measurements with this device are done at 25 °C (77 °F). For instance, suppose that one has 350 mL of beer with 20 mL of headspace and suppose that the Zahm and Nagle instrument reads

$$V_{air} = 3 \text{ mL of air in headspace}$$

and

$$P_{total} = P_{headspace} = 36 \text{ psi.}$$

Table 6.3 shows that, ignoring air corrections, this pressure yields 2.52 volumes of CO_2. The measured pressure $P_{total} = 36$ psi is actually the sum of the partial pressure of CO_2 and the partial pressure of air. By using both Boyle's and Dalton's laws, the partial pressure of air can be computed as follows:

$$P_{air \text{ in headspace}} = (V_{air \text{ at atmospheric pressure}} / V_{headspace}) \times P_{atmospheric \text{ pressure}}$$

$$= [(3 \text{ mL of air})/(20 \text{ mL of air plus } CO_2)] \, (1 \text{ atm}) = 0.15 \text{ atm}$$
$$= (0.15 \text{ atm}) \, (14.7 \text{ psi})/(1 \text{ atm}) = 2.2 \text{ psi} \quad.$$

Thus, the partial pressure due to CO_2 is actually

$$P_{CO_2} = 36 \text{ psi} - 2.2 \text{ psi} = 33.8 \text{ psi,}$$

and interpolating from Table 6.3 at 25 °C gives

$$CO_2 \text{ volumes} = 2.37 \text{ Vol.} + \frac{(33.8 \text{ psi} - 33 \text{ psi})}{(34 \text{ psi } - 33 \text{ psi })} \times (2.42 \text{ Vol.} - 2.37 \text{ Vol.})$$

$$= 2.41 \text{ Vol.}$$

Once the headspace air volumes fall below 1 mL, it is customary to ignore such corrections (ASBC, 1987).

DISPENSING DRAFT BEER

As a final application of the gas laws, consider the proper dispensing of carbonated bottled beer. There are three crucial variables:

1. Equilibrium pressure (P_E). This variable is the external pressure required to maintain the beer's gas content. In the numerical example given next, the beer contains 2.5 vols. of CO_2 (no air for this example) and is at 4.5 °C (about 40 °F). Table 6.3 shows that P_E is approximately 12 psi in this case.

2. Gravity head (P_H). This variable is the pressure to push the beer against gravity from the beer tank to the tap. A widely used rule is that it takes 0.042 psi/in., or equivalently, 1 psi is required to lift beer 2 ft. vertically (Miller, 1992; Wheeler, 1998).

3. Line resistance (P_{LR}). This variable is the pressure required to overcome the resistance in the tubing leading from the beer keg to the tap. For either vinyl or polyethylene hose, a 3/8 in. line provides a resistance pressure of 0.1 psi/ft. whereas lines of 1/4 in. and 3/16 in. diameter require 0.6 and 2.2 psi/ft., respectively (Wheeler, 1998).

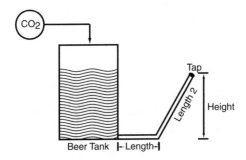

Beer-dispensing system

The effective pressure is defined as

$$P_{eff} = P_E - P_H - P_{LR}.$$

To prevent overflowing and gushing, the line length is chosen so that the effective pressure is zero ($P_{eff} = 0$). Then

$$P_E = P_H + P_{LR}.$$

This relationship results in a balanced system, and if the exogenous CO_2 supply is held at P_E, then the system would be in equilibrium. There would be no flow, but at the same time there would be no loss of CO_2. A good working rule is that if the supplied CO_2 pressure exceeds P_E by 2 psi in a balanced system, then a flow rate of 1–1.2 gal./min (2.1–2.6 oz./s) will result. This flow rate is considered ideal; there is no loss of CO_2 during dispensing (Miller, 1992; Wheeler, 1998).

To give a numerical example, consider the case of a typical draft box (such as one from the Superior Products Catalog, St. Paul, Minnesota).

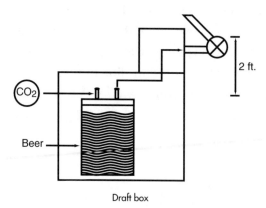

Draft box

Suppose that the beer has 2.5 vols. of CO_2 and is dispensed at 4.5 °C (about 40 °F). As noted previously, this arrangement gives $P_E = 12$ psi. The gravity head is 2 ft. so,

$$P_H = 0.042 \times 2 \times 12 = 1.$$

Thus, to have a balanced system, one needs

$$P_{LR} = P_E - P_H = 12 - 1.0 = 11.0 .$$

Suppose a 3/16 in. tubing is used. Then the line length required is

$$11/2.2 = 5 \text{ ft.}$$

As noted above, the desired dispensing pressure to achieve balance is

$$P_E + 2 = 12 + 2 = 14 \text{ psi .}$$

This analysis does not take into account the possible CO_2 pickup in the beer over time by holding 14 psi top pressure on it. This additional CO_2 will lead to an unbalanced system—and hence to excess foaming at the tap—in addition to the undesirable increase in beer CO_2 volumes. An alternative is to use a CO_2 and N_2 mix for the driving pressure. The idea is to have sufficient CO_2 to balance the equilibrium pressure of beer and to let the N_2 overcome the gravity head and line resistance.

As a specific example, suppose that a 75% N_2 and 25% CO_2 mix is used. In such a mix, the pressure due to CO_2 will be roughly 1/3 that of N_2. Thus, the desired top pressure is as follows:

CO_2 pressure	12 psi
N_2 pressure	36 psi
Top pressure	48 psi

This top pressure may seem high at first glance, but it must be kept in mind that the solubility of N_2 in beer is a factor of 80 less than CO_2. Other numerical examples are in Taylor et al. (1992).

The use of N_2 and CO_2 mix for carbonating beer is also discussed in Taylor et al. (1992). This approach can be very tricky owing to high gas losses during processing. If these can be managed, however, the result can be a beautiful and highly stable foam stand. This can be explained by partial pressures of the individual gases. First, consider the previous example in which the equilibrium pressure is 12 psi at 40 °F. A typical foam bubble can be described schematically as follows:

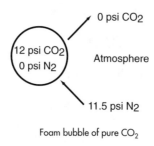

Foam bubble of pure CO_2

That is, there is a large partial-pressure differential with the ambient environment leading to CO_2 escape and collapsing bubbles. Now consider a 75% N_2 and 25% CO_2 mix of at 16 psi:

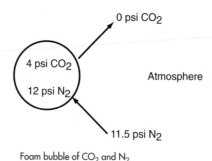

Foam bubble of CO_2 and N_2

The nitrogen is nearly in equilibrium with the ambient environment, whereas the pressure differential for CO_2 has been reduced by a factor of 3. In this situation, the gas loss is considerably less than in the first example and hence provides more support for the foam stand.

I would like to conclude this chapter and indeed this book on a personal note. For me, the dispensing of a new brew is an event of great significance. The beer in the glass is a product that represents a long list of decisions and actions on my part. First of all, there is the choice of brewing materials and the way they are processed. Actually, of greater significance is my choice of yeasts and what I have done to see that they were in appropriate condition to do whatever it is I want them to do. Of even greater significance are the actions I have taken to protect my concoction from adversaries, be they foreign microbes or oxygen. All of this is in that glass and represents the deepest joy of brewing.

References

Amerine, M. A., H. W. Berg, and W. V. Cruess. 1972. *Technology of Wine-Making*. Westport, CT: AVI Publishing.

Amerine, M. A., and M. A. Joslyn. 1973. *Table Wines*. Berkeley: Univ. of California Press.

Anners, B. J., and C. W. Bamforth. 1982. *J. Inst. Brew.* 88.

Arcay-Ledezma, G. J., and J. C. Slaughter. 1984. *J. Inst. Brew.* 90.

Aries, V., and B. H. Kirsop. 1977. *J. Inst. Brew.* 83.

Arnold, J. P. 1911. *Origin and History of Beer and Brewing*. Chicago: Wahl-Henius Inst.

Axcell, B., and P. Torline. 1998. *MBAA Tech. Qr.* 28.

Bach, J., and F. Fersing. 1997. *Brauwelt* 15:3.

Bailey, K. 1998. *The Life of Webern*. Cambridge: Cambridge Univ. Press.

Bamforth, C. W. 1999. *Brauwelt* 17(2).

Bamforth, C. 1998. Beer: *Tap into the Art and Science of Brewing*. New York: Plenum.

Barker, R. L. 1983. *J. Inst. Of Brew.*

Beran, K., E. Streiblova, and J. Lieblova. 1966. Bratislava: Proc. of the 2nd Symposium on Yeasts.

Berg, K. A.. 1991. *MBAA Tech. Qr.* 28.

Bickham, S. 1998. Focus and Flavor. *Brewing Techniques* 6(3).

Blenkinsop, P. 1991. *MBBA Tech. Qr.* 28:3.

Bourne, D. T., T. Powleslarrd, and R. E. Wheeler. 1982. *J. Inst. Brew.* 88.

Buckee, G. K., P. T. Malcolm, and T. L. Peppard. 1982. *J. Inst. Brew.* 88.

Cahill, G., P. K. Walsh, and D. Donnelly. 1999. *J. Am. Soc. Brew Chem.* 57:2.

Cantrell, I. C., and D. L. Griggs. 1996. *MBAA Tech. Qr.* 33:2.

Chapon, L. 1981. *Brewing Science*, vol. 2.

Comrie, A. A. D. 1967. *J. Inst. Brew.* 73.

Conn, E. E., and P. K. Stumpf. 1976. *Outlines in Biochemistry*. New York: John Wiley.

Dadie, M. 1980 *Brewers Digest* (April).

Dalgliesch, C. D. 1977. Amsterdam: Proc. of the 16th *Eur. Brew. Conv. Cong.*

De Clerck, J. A. 1957. *A Textbook of Brewing*, 2 vols. London: Chapman and Hall.

Diefenbach, M. J. 1996. *Brauwelt* 14:4.

Doderer, A., et al. 1991. *Proc. Eur. Brew. Conv. Cong.* 23.

Engan, S. 1991. *Brauwelt* 9.

Fix, G. J. 1986. *Beer and Brewing*. Trans. from the Natl. Conf. on Quality Beer and Brewing, vol. 6. Boulder: Brewers Pubications.

————. 1987. *Beer and Brewing*. Trans. from the Natl. Conf. on Quality Beer and Brewing, vol. 7. Boulder: Brewers Publications.

————. 1990. *Beer and Brewing*. Trans. from the Natl. Conf. on Quality Beer and Brewing, vol. 10. Boulder: Brewers Publications..

————. 1992. *Zymurgy* 15:3-5.

————. 1993. *Brewing Techniques* 1:2.

————. 1994. *Zymurgy* 17.

————. 1998. *Brewing Techniques* 6:6.

Fix, G.J. and L.A. Fix. 1997. *An Analysis of Brewing Techniques*. Boulder: Brewers Publications.

————. 1999. Texas Brew. Assoc. Newsletter (June).

Forster, C., L. Narziss, and W. Back. 1998. *MBAA Tech. Qr.* 35:2.

Fugelsang, K. C., M. M. Osborn, and C. J. Muller. 1993. In *Beer and Wine Production*. Edited by B. H. Gump. Washington, D.C.: Am. Chem. Soc. Sym. Series 536.

Garetz, M. 1994. *Using Hops*. Danville, CA: Hop Tech. Publications.

Gibson, R. M., and C. W. Bamforth. 1982. *J. Inst. Brew.* 89.

Guinard, J. X. 1990. *Lambic*. Boulder: Brewers Publications.

Haley, J., and T. L. Peppard. 1983. *J. Inst. Brew.* 89.

Hardwick, W. A. 1978. *Brewers Digest*. Oct.

Harrison, J. G., et al. 1963. *J. Inst. Brew.* 69.

Hashimoto, N. 1981. In *Brewing Science*, vol. 2. Edited by J.R.A. Pollock. New York: Academic Press.

————. 1986. *Brewers Digest* (June).

Hashimoto, N., and Y. Kuroiwa. 1975. Kirin Brew. Co. Research Report 18.

Haunold, A. 1998. *The New Brewer* 15:6.

Heath, H. B. 1988. *Flavor Chemistry and Technology*. New York.: AVI-Von Nostrand Reinhold.

Herrmann, H., et al. 1985. *Brauwelt* 3(1).

Hind, H. L. 1950. *Brewing: Science and Practice*. London: John Wiley.

Holtermand, A. 1963. Recent Chemical Aspects of the Browning Reaction. *Proc. Euro. Brew. Conv. Cong.*

Hough, J. S., et al. 1981. *Malting and Brewing Science*, 2 vols. London: Chapman and Hall.

Hughes-Hallet, P. et al. 1998. *Calculus*. New York: John Wiley.

Huige, N. J. 1993. In *Beer and Wine Productions*. Washington, DC.: Amer. Chem. Soc. Sym. 536.

Ingledew, W. M. 1979. *J. Am. Soc. of Brew. Chem.* 37.

Ishibashi, Y., et al.. 1997. *J. Am. Soc. Brew. Chem.* 55:1.

Jackson, M. 1997. *Pocket Guide to Beer*. New York: Simon and Schuster.

Kieninger, H. 1977. *J. Inst. Brew.* 83.

Knudsen, F. B. 1977. In *The Practical Brewer*. Edited by H.M. Broderick. Madison: MBAA.

Kolbach, P. 1960. *Monatsschrift fŸr Brauerei* 13.

Krauss, G., H. Waller, and R. Schmid. 1955. *Brauwelt*.

Kunze, W. 1996. *Technology Brewing and Malting*. Berlin: Verlagsabteilung.

Lee, S. Y., et al. 1980. *J. Inst. Brew.* 86.

Lermusuary, G., et al. 1999. *J. Am. Soc. Brew. Chem.* 57:1.

Lewis, D. 1998. *The New Brewer* (Jan./Feb.)

Lewis, M. J. 1963. *Proc. Am. Soc. Br. Chem.*

Lewis, M. J., and T. Young. 1995. *Brewing*. London: Chapman and Hall.

Lieberman, C. E. 1980. *Brewers Digest* (Dec.)

———. 1984. *Brewers Digest* (Dec.)

Linemann, A., and E. Krueger. 1997. *Brauwelt* 15:4.

MacFarlene, W. D. 1970. *Industry Sponsored Research on Brewing*. Chicago: Brewing Industry Research Inst.

McMurrough, I., D. Madigan, and R. J. Kelley. 1997. *J. Am. Soc. Brew. Chem.* 55:2.

Meilgaard, M. C. 1975. *MBAA Tech. Qr.* 12:3.

———. 1991. *MBAA Tech. Qr.* 21.

———. 1977. In *Practical Brewer*. Edited by H. M. Broderick. Madison: MBAA.

Melm, G., P. Tung, and A. Pringle. 1995. *MBAA Tech. Qr.* 32:1.

Methods of Analysis. 1987. 8th rev. ed. St. Paul: Am. Soc. Brew. Chem.

Miller, D. 1992. *Beer and Brewing*. Trans. from the Natl. Conf. on Quality Beer and Brewing 12. Boulder: Brewers Publications.

Mitter, M. 1995. *Brauwelt* 13:4.

Moll, M. M. 1995. Water. In *Handbook of Brewing*. Edited by W. A. Harwick. New York: Marcel Dekker.

Morrison, N. M., and D. S. Bendrak. 1987. *MBAA Tech. Qr.* 24.

Murrary, C. R., T. Barich, and D. Taylor. 1984. *MBAA Tech. Qr.* 21.

Nakatani, K., et al. 1984. *MBAA Tech. Qr.* 21.

Narziss, L. 1989. *Brauwelt* 7.

———. 1990. *Brauwelt* 8.

———. 1992. *Brauwelt* 10:4.

———. 1997. *Brauwelt* 15:1.

———. 1998. *Brauwelt* 16:1.

Narziss, L., H. Miedaner, and A. Gresser. 1984. *Brauwelt* 2.

Neve, R. A. 1991. *Hops*. London: Chapman and Hall.

Nugy, A. L. 1937. *Brewers Manual*. Jersey Publication Co.

Nyborg, M., H. Outtrup, and T. Dreyer. 1999. J. *Am. Soc. Brew. Chem.* 57:1.

Oliver, F., and B. Dauman. 1988. *Brauwelt*. 6:3-4.

O'Rourke, T. 1996. *Brewers Digest* (Oct.).

Owades, J.L. 1981. *MBAA Tech. Qr.* 18.

———. 1985. Lecture notes. San Francisco: Center for Brewing Studies.

Owades, J. and M. Plam. 1988. *MBAA Tech. Qr.* 25:4.

Owades, J. L., and J. Jakovac. 1959. Proc. Am. Soc. Br. Chem.

Panchal, C. J., and G. G. Stewart. 1980. *J. Inst. Brew.* 86.

Pasteur, L. 1876. *Etudes sur la Bière*. Paris: Gauthier-Villars.

Peacock, V. E. 1992. *MBAA Tech. Qr.* 29:3.

Peacock, V. E., and M. L. Deimzer. 1981. *J Am Soc. Br. Chem.* 39(4).

Peacock, V.E., et al. 1981. *J. Agric. Food Chem.* 29.

Piendl, A. 1970-1990. Biere aus Aller Welt. *Brauindustrie.*

Pierces, J. S. 1982. *J. Inst. Brew.* 88.

Pipes, J. U. 1978. *MBAA Tech. Qr.* 15:1.

Pogh, T. A., J. M. Maurer, and A. T. Pringle. 1997. *MBAA Tech. Qr.* 34:3.

Prechtl, C. 1964. *MBAA Tech. Qr.* 4:1.

Preis, P., and M. Mitter. 1995. *Brauwelt* 13:4.

Quain, D. E., and K. S. Tubb. 1982. *MBAA Tech. Qr.* 19.

Quain, D. E., P. A. Thurston, and R. S. Tubb. 1981. *J. Inst. Brew.* 87.

Reed, G., and T. Nagodawithana. 1991. *Yeast Technology*. Westport. CT.: AVI Publishing.

Rehberger, A. J., and G. E. Luther. 1995. Brewing. In *Handbook of Brewing*. Edited by W. A. Hardwick. New York: Marcel Dekker.

Rigby, L. 1972. *J. Am. Soc. Brew. Chem.* 30.

Russell, Inge. 1995. Yeast. In *Handbook of Brewing*. Edited by W. A. Hardwick. New York: Marcel Dekker.

Ryder D. et al. 1988. *MBAA Tech. Qr.* 25.

Ryder, D. S., and J. Power. 1995. Miscellaneous Ingredients in Aid of the Process. In *Handbook of Brewing*. Edited by W. A. Hardwick. New York: Marcel Dekker.

Saltukoglu, A., and J. C. Slaughter. 1983. *J. Inst. Brew.* 89.

Scheer, F. 1988. *American Brewer* 3:38.

———. 1990. *The New Brewer* (July/Aug.).

Schneider, J. 1997. *Brauwelt* 15:3.

Seaton, J. C., A. Suggett, and M. Moir. 1981. *MBAA Tech. Qr.* 18:1.

Sebree, B. R. 1997. *MBAA Tech. Qr.* 34:3.

Shannon, R.V.R., G.D. John, and A. M. Davis. 1978. *Brewers Digest* (Sept.).

Siebert, K. J., et al. 1986. *MBAA Tech. Qr.* 23.

Siebert, K. J., and P. Y. Lynn. 1997. *J. Am. Soc. Brew. Chem.* 55:2.

Siro, M. R., and T. Lovgrem. 1980. *Brewers Digest* (Sept.).

Stewart, C. G., C. J. Panchal, and I. Russell. 1983. *J. Inst. Brew.* 89.

Stucky, G. J., and M. R. McDaniel. 1997. *J. Am. Soc. Brew. Chem.* 55:1.

Taylor D. G., et al. 1992. *MBAA Tech. Qr.* 29.

Taylor, D. G. 1990. *MBAA Tech. Qr.* 27.

Thomas, D. 1986. *Beer and Brewing*. Trans. from the Nat. Conf. on Quality Beer and Brewing. Boulder: Brewers Publication.

U. S. Patent No. 5,304,384. Labatt Brew. Co., April 19, 1994.

Verzeli, M., and D. de Keukeleire. 1991. *Chemistry and Analysis of Hop and Beer Bitter Acids*. Amsterdam: Elsevier.

Vincenti, W. G., and C. H. Kruger. 1982. *Introduction to Physical Gas Dynamics*. New York: Krieger.

Wackerbauer, K. 1993. *Brauwelt* 11:3.

Wahl, R., and M. Henius. 1908. *American Handbook of Brewing*. Chicago: Wahl-Henius Institute.

Wainwright, T.. 1973. *J. Inst. Brew*. 79.

White, F. H. 1977. *Brewers Digest* (May).

Zangrando, T. 1979. *Brewers Digest* (April).

Zimmermann, A. 1904. *Brauereibebriebslehre*. New York.

INDEX

α-acids, in hops resins, 54–55
α-amylase
    action in mashing, 40–41
    as liquefaction enzyme, 46
α/β ratios in hops, 56–57
Abuse in trade, 142
Acetic acid bacteria, 122–123
Acid-forming enzymes in malting and mashing, 38
Acidification
    of brewing water, 4–5, 8–9
    by direct mash *versus* salt additions, 13
Acid rest, 46
Acrospire, growth of, 37, *37*
Additives covering staling effects, 127
Africa, use of malted sorghum, 45
α-glucans (amylopectins), 18–20, 40–41
Aldehydes, staling of, 135
Ales, dimethyl sulfide levels of, 33
Alkaline water
    boiling of, 10–11
    treatment residual, 12
    treatment with lime, 10–11, *11*
Amadori rearrangement, during kilning, 43–44
American Society of Brewing Chemists (ASBC)
    °ASBC malt color units, 41–42, *42*
    measurement method of beer stability, 143, *143,* 144
    study of extract and ethanol levels, 93
Amines, effects on beer flavor, 36
Amino acids
    assimilation rate in fermentation, 24, 25, *25*
    classification of, 24–25, *25*
    free amino nitrogen (FAN) content in wort, 26
    percentage absorbed by yeast in fermentation, 100–101
    as proteins building blocks, 23–29
    as source of nitrogen in brewing, 23–29
Amylopectins. See α-glucans
Amylose sugars, 17–18
Anabolic reactions, definition of, 94
Anti-fungal spraying source of sulfury flavors, 64–65

Antioxidants, natural source in iso—acids, 136–137
Arabinose, 21
Aromatic rings, role in brewing, 30–32
Astringent compounds
    extraction during sparging, 51–52
    materials in husks, 45–46
Attenuation
    apparent/real, 48
    behavior of yeasts, 81–82, *83*
Auto-oxidation mechanism in beer staling, 130–131
Avogadro's number, 158

β-acids in hops, 54–57
Bacteria
    in fermentation, 119–120, *120,* 121, *121,* 122, *122,* 123, *123,* 124, *124,* 125, *125*
    heterofermentative, 121
    homofermative, 122
    infection as source of dimethyl sulfide, 34–35
Baeyer condensation, 136
Balling, Carl Joseph, study of extract and ethanol levels, 92
β-amylase, action during malting, 40–41
Barley
    β-glucans level, 21, 22
    germinating, *37*
    moisture uptake in kernels, 1
    raw, *37*
B-complex vitamins, sources during fermentation, 35–36
β-D-glucose, 16
Beer
    freshness and role of phenolic compounds, 31–32
    phenolic compounds in oxidation, 31–32
    smoothness of, 51
    sour types, 45, 122
    stability of foam, 48–49, 144, 146, 171–173
    staling by lipid-active enzymes, 43
    staling defects, 127–128, *128,* 129–139
    "white" type, 50–51
Beer, staling of. *See* Staling of beer

Cross-flow cartridge filtration, 149–150
CSA. *See* Cold-side aeration
Curing phase, for malt color, 42–43, *43*
Currant aroma in staling of beer, 128
Cytolytic modification, measurements of, 38–39
Czech Pilseners, mineral content of, 15

Dalton's law of partial pressures, 162
Dark beers
  high sulfate levels, 14
  synergism with carbonates, 14
Dextrins, 18–20, 22
Diacetyl production during fermentation, 113–114, *114,* 115–116, *116,* 117, *117,* 118–119
Diastatic power (DP) of malt enzyme system, 42
Dimethyl sulfide (DMS), 32–35
  in fermentation, 123, *123,* 124, *124,* 125, *125*
  formation during malting, 44–45
  levels in finished beers, 33–34
  precursor conversion, 45
  production/reduction in wort boiling, 71–75
  "scrubbing out," 34, 125
Dimethyl sulfoxide (DMSO)
  as dimethyl sulfide precursor, 34–35
  in fermentation, 125, *125*
Dispensing draft beer, 169–173
Dissociation of ions, 3, 4
Distillation in "ice beer" process, 150–151, *151*
DMS. *See* dimethyl sulfide
DMSO. *See* dimethyl sulfoxide
Dortmunder Actien-Export, mineral content of, 14–15
Dortmund water treatment techniques, 14–15
Draft beer, dispensing of, 169–173
Drying phase for malt color, 42–43, *43*

Ehrlich mechanism of fusel alcohol production, 107, *107*
"Elegant" flavor, humulene source of, 58
Embden-Meyerhof-Parras (EMP) pathway, xii
Endosperm exposure during milling, 45–46
Entner-Doudoroff (ED) pathway in fermentation, 124, *124,* 125
Enzymes
  crucial to malting and mashing, 38–39
  deactivation during curing, 42
  role in brewing, 28–29

Essential oils in hops
  components of, 57–59, *60, 61, 62,* 63–65
  hydrocarbon components, 57–59
Esters, flavor sources in fermentation, 111–112, *112,* 113
European Brewery Convention beer-haze units (EBC), 143, *143*
European Brewing Congress /EBC malt color units, 41–42, *42*

FAN. *See free amino nitrogen* (FAN)
Farnesene in hops, 58, 59
Fatty acids
  role in brewing, 29–30
  role in staling of beer, 132–135
  saturated types from yeast metabolism, 109, *110,* 111
Fermentation
  anaerobic or *Pasteur effect,* 93
  bacterial pathogens, 119–120
    acetic acid bacteria, 122–123
    lactic acid group, 120, 121, *121,* 122, *122*
    sulfur-producing bacteria, 123, *123,* 124, *124,* 125, *125*
  Crabtree effect of respiration inhibition, 93–94
  EMP pathway (glycolysis)
    carbon splitting, 103
    formation of pyruvic acid, 91, 105
    Gay-Lussac formula, 91
    phosphorylation of glucose, 101–103
    redox reactions of, 104
  importance of, xii–xiv
  initial period
    acetyl coenzyme A, 91
    amino acid intake by yeast cells, 99–101
    formation of pyruvic acid, 91
    glycogen content of pitching yeasts, 97, 98
    internal energy yeast cell reserves, 97
    maltose inhibition, 90
    "shock excretion" of yeast cell walls, 90, 100
    sterol synthesis for yeast cells, 89, 95–97, 98
    Strickland reaction, 99–100
    sugar intake by yeast cells, 98, *98,* 99
    uptake of oxygen, 95–96
    wort trub stimulation of yeast metabolism, 96–97
    yeast growth, 90–91
  minor pathways
    diacetyl production, 113–114, *114,* 115–116, *116,* 117, *117,* 118–119

Linoleic acid, 29–30
°Linter (°L) units, 42
Lipid-active enzymes in beer staling, 43
Lipids, role in brewing, 29–30
Liquor:grist ratio, 49
London
    brewing water composition, 13
    dark beers, 14
°Lovibond malt color units, 42

Magnesium
    role in fermentation, 5
    yeast requirements for, 36
Maillard products, xvi
    and dark malts, 14
    formation during wort boiling, 44
    formed during kilning, 43–44
    roasted malt concentrations, 44
Maillard reactions, 2, 36
    role of amino acids in malt drying/kettle
        boil, 23–24
    in wort boiling, 75–78
Malt, xvi
    for brewing, 45
    carbohydrate constituents, 15–36
    color of, 41–42, *42*, 43, *43*, 44
    diastatic power (DP), 42
    drying (kilning) of, xvi
    Maillard products, 14
    milling of, 45–46
    as protein source in brewing, 23–29
    pyrazines/pyrroles in dark malts, 44
    storage over time, 41
    temperatures for color control, 42–43, *43*
Malt drying, role of amino acids, 23–24
Malting
    approximate color correspondences, *42*
    breakdown of proteins, 28–29
    carbohydrate modification of, 39, *39*,
        40–41
    color and temperature data, *43*
    cytolysis, 37, 38–39
    definition of, 36
    enzyme development of, 36, 38
    formation of sulfur compounds, 44–45
    hormone release, 38
    Maillard products, 37, 44
    and mashing, 45–47, *47*, 48, *48*, 49–50, *50*,
        51–52
    modification of proteins and starches, 37
    moisture level of, 41
    starch contents of, *39*
    steeping, 37, *37*

termination of process in kilning, 41
Maltose as disaccharide, 17
Maltotriose as trisaccharide, 17
Mash
    acidification *versus* salt additions, 12–13
    pH of, 8, 49–50, *50*, 51
    preparation of main mash, 50
    preparation of sour mash, 50
    separation from the grain, 51–52
Mash Filter 2001, 46
Mashing
    key parameters for systems, 46–47, *47*
    lactobacilli acidification of, 122
    pH establishment of, 46
    process of, 45–47, *47*, 48, *48*, 49–50, *50*,
        51–52
    rests, ranges of, 46–47, *47*
    thickness, description of, 49
Megasphaeria gram-negative bacteria in fer-
        mentation, 123–124
Melanoidin-induced oxidation in staling of
        beer, 134–135
Melanoidins, 36
    formation during kilning, 43–44
    as wort reductones, 76
Melanoids, xvi
Mercaptan sulfur compounds, 35
Metabolic pathways. *See* Fermentation, EMP
        and minor pathways
Moisture
    level in malt, 41
    uptake in barley kernels, 1
Molarity of gas in solution, defined, 163
Mole, definition of, 6
Monosaccharide, glucose classification, 17
Morphology of yeasts, 80
Munich
    brewing water composition, 13
    dark beers, 14
    nonenzymatic browning of malts, 44

NEB. *See* Nonenzymatic browning (NEB)
Nitrogen compounds, 23–29
"Nonadditive brewing," 127, 149–150, *150*, 151,
        151, 152
Noncultured yeast, xiii
Nonenzymatic browning (NEB), 43–44
    in wort boiling, 75–78
Nonflocculating strain of yeast, 80–81
Nonreducing carbohydrates, 22
North America dimethyl sulfide levels in ales,
        33, 34
Nucleic acids, effects on beer flavor, 36

Wine yeasts, 85–86
°W-K
    See Windisch-Kolbach units, 42–43
Wort
    amino acids assimilated by yeast in fermentation, 100–101
    attenuation in yeast flocculation, 80–81
    carbohydrate content of, 26
    chilling of, 28
    clarity of, 30
    "cold break," 78
    coliform bacteria spoilers, 123, *123*, 124, *124*, 125, *125*
    composition of, xiii
    evaporation rates and flavors, 77–78
    free amino nitrogen (FAN) amino acid content of, 26
    importance of carbonate/bicarbonate, 5, 7
    nitrogen levels of, 28–29
    oxygenation of, xii
    production/reduction of dimethyl sulfide (DMS), 71–75
    protein coagulation "hot break," 78
    proteins, content of, 23
    pyrroles as flavor source, 77
Wort, boiling of
    hops, resins and essential oils, 53–59
    isomerization of hop compounds, 65–67
    oxidation/polymerization of hop resins and oils, 67–69, *69*, 70, *70*, 71
    oxygen-bearing components, 59, 63–64
    selected hops data, *60, 61, 62*
    sulfur-containing compounds, 64–65
W-308 yeast strain, 84

Xylose, 21

Yeast-active fining agents, 148, *149*
Yeasts
    behavioral characteristics of, 80–83, *83*, 84
    birth-scar theory of multiplication, 86–89
    "bottom cropping" of, 81
    care and feeding of brewing strains, xii, 86–89
    classification of, 79–80
    contaminated types, 84–86
    ester signature flavors, 111–112, *112*, 113
    individual strains, xiii
    metabolism and phenolic compounds, 32
    number of yeast cells, xii
    oxygen demands, 81–82
    physiological condition of, xii, xiii
    pitching considerations, 81–83, *83*
    preferred temperature ranges, 82–83

reusing of, 86–89
"rousing" of, 81
sensitivity to alcohol, 81
strains affected by the *Crabtree effect,* 94
sugar-fermenting abilities, 81–83, *83*
super attenuating types, 84–85
wine yeasts, 85–86

Zinc, yeast requirements for, 36
Zymomonas bacteria in fermentation, 124

ABOUT THE AUTHOR

The late George Fix was a highly respected name in the world of brewing. He authored four books published by Brewers Publications and scores of technical articles. An avid and dedicated homebrewer, he won numerous awards, including the Ninkasi Award from the American Homebrewers Association. He traveled widely to lecture and consult and he was a dynamic, entertaining and much anticipated speaker at brewing conferences and events.

Dr. Fix was also a highly respected mathematician. He held a Ph.D. in Mathematics from Harvard University and a Master's Degree from Rice University. Following thirteen years as Professor of Mathematics at the University of Texas at Arlington, he was Professor and Chairman, Department of Mathematics at Clemson University, Clemson, South Carolina.

During his lifetime, Dr. Fix was a member of the American Mathematical Society, American Society of Brewing Chemists, Master Brewers Association of America and American Association for the Advancement of Science. Dr. Fix is survived by his wife Laurie who co-authored many of his brewing publications.